THE HOCHMAN
ENCYCLOPEDIA
of American Playing Cards

Tom Dawson

Judy Dawson

116/500

THE HOCHMAN
ENCYCLOPEDIA
of American Playing Cards

TOM and JUDY DAWSON

U.S. GAMES SYSTEMS, INC.
Publishers
Stamford, CT 06902 USA

Printed in the United States of America

Hardcover: ISBN 1-57281-297-4
Softcover: ISBN 1-57281-299-0

©2000 U.S. Games Systems, Inc.
Stamford, CT 06902 USA

First printing 2000

Library of Congress Cataloging-in-Publication Data

Dawson, Tom, 1936-
    The Hochman encyclopedia of American playing cards / by Tom and Judy Dawson.
      p. cm.
    ISBN 1-57281-297-4
    1. Playing cards--United States--History. 2. Playing card makers--United
Stated--History. I. Title: encyclopedia of American playing cards. II. Hochman, Gene.
III. Dawson, Judy. IV. Title.

GV1234.U6 D38 2000
688.7'54'0973--dc21

TABLE OF CONTENTS

PREFACE

The origins of this book go back to 1976 when Gene Hochman published Part I and Part II of the Encyclopedia of American Playing Cards. These two volumes constituted a comprehensive, illustrated listing of certain categories of known non-standard American playing card decks. Gene went on to publish Part III in 1978, Part IV in 1979 and Parts V and VI in 1980 and 1981 respectively, these latter two adding many listings to the categories covered in the first four volumes.

These books were a prodigious effort, requiring a huge amount of research, and many, many hours of typing, photographing and compilation – all in days when the technology for this type of project was a far cry from that available today.

In Part I, Gene wrote a simple preface which we would like to quote from in order to give the reader a flavor of what was going on in his mind as he started the work.

"When fascinated by the first deck of antique playing cards that came into my possession many years ago, little did I know that some day I might actually compile a book on the subject. It is surely not the first work of this kind, for in my own library, I have many fine books by many fine authors without which my collecting of playing cards would have been an impossible task.

This will be, however, the FIRST book devoted exclusively to AMERICAN playing cards. The subject is vast. In compiling Part 1 of this "Encyclopedia of American Playing Cards", several nearly finished volumes became prematurely outdated. Outdated not by NEW ISSUES, but by NEW DISCOVERIES. One must start somewhere! No doubt, I will have learned of omitted decks before this edition reaches the press. But, the foundation will have been laid. The readers and I will be able to build and improve our knowledge with greater ease with something available to use as a guide.

The complete Encyclopedia will be issued in four sections. The research involved with each portion might never be completed if I tried to retain all of the information to issue at one time.

PART 1 *will consist of Souvenir Cards, Exposition & World's Fair Issues, Cards from the Entertainment and Sports Field, War Cards, Railroad Cards and Political & Patriotic Editions.*

PART 2 *will consist of Standard Decks from the earliest makers to pre World War II. I will endeavor to make this as complete as possible paying particular attention to new innovations as they were introduced, such as indices, etc.*

PART 3 *will consist of Transformation Cards and Tobacco Insert Playing Cards. These categories are by far the most interesting of any to be discussed. Details in this part must be carefully noted and many of the decks must be photographed in their entirety.*

PART 4 *will consist of all other subjects not covered in the above three. Advertising decks, No Revoke decks, New Suit signs, Fortune Telling, Instructional decks, and those of Odd Shapes and made of Odd Materials will be among those covered.*

An important feature of every Part, as well as the entire Encyclopedia, is that almost every deck will be identified with a photograph. No longer will collectors be frustrated in their identity search with that often none too accurate description."

Those readers who are familiar with Gene's original volumes will recognize that he changed his mind as to the order of his publication and his contemplated Part 3 on transformation and insert cards actually was published as Part II. Similarly, his grouping for Part 4 became Part III and he finished with a very detailed Part IV covering the standard decks of all the known manufacturers. But let's return to some of his earlier commentary.

Gene decided to start with various categories of souvenir type decks because he thought they were already quite collectible and familiar and because the number of listings was finite, or nearly so. The thoroughness of his work is evident in the relatively small number of additions to the Part I categories in Parts V and VI and, indeed, in this much later rewrite. In Gene's words:

"It may seem unusual to start Part 1 with the subject of Souvenir playing cards. The reason is obvious! While certainly not the earliest, Souvenir cards are just as surely the most plentiful of all the 'collectible' editions. It is not at all unusual to find a deck of these cards, issued in the 19th century, in mint condition, opened but never used for playing. Unlike early decks bought for playing, these cards were usually purchased as keepsakes. Due to having pictures on each card, they might have been often shown, but never 'used'.

Souvenir cards cover many categories. There are those issued at Expositions and Fairs, those to commemorate

Political or Sporting events, as well as Railroad decks used to promote the various lines or purchased to be kept as a souvenir of having travelled them. A great many decks were issued by States, by Cities, or even by National Parks or other places of interest. Then, as well as now, keepsakes were sought by most everyone.

Despite years of research, one could never be sure to list the perhaps thousands of different decks in every souvenir category. Even today, one can visit an obscure winery and find that for one dollar they can purchase a deck of cards to remember this visit. To therefore limit the listing of cards in this publication to the most collectible, we will start at 1876 at the Centennial Exposition and end at the New York World's Fair of 1939. We will also restrict the state, city, parks and railroad decks to those which carried a different photo on the face of each card. This latter category will be limited to the wide editions and cover none of the Bridge Size decks which have been published so plentifully during the last thirty-five years."

In 1989, after Gene and Ellen had retired to Florida, he decided it was time to update his Encyclopedia. For many months he labored away, taking pictures of new listings discovered since Part VI was published in 1981, and updating material in his original volumes for information which had become known, or clearer, since that time. However, he found it was eating into the time he relished for other activities that had become part of his new way of life in the warmer climes of the south. For instance he took up golf again after an absence of many years and he returned to high level duplicate bridge play, an avocation where he had long before achieved Life Master ranking, playing as many as four or five times a week.

Of course he also continued adding to his wonderful playing card collection while continuing to run his playing card business dealing in new issues and his Encyclopedia and running his quarterly auctions of old, unusual and rare playing cards and related items. As we seemed to share many of the same passions as Gene, we tried to get to Boynton Beach whenever we could slip away from the winter weather in the north and enjoy some golf, some duplicate bridge, some card stories (Gene had a seemingly endless supply) and some happy, relaxing time with him and Ellen at their home.

It was on one of these occasions, that Gene asked us if we would help him finish the rewrite of the Encyclopedia, do some traveling to visit collectors and photograph hard to find decks, and arrange the publishing with old friend Stuart Kaplan of U.S. Games. This did not seem such a daunting task and we quickly said we would be glad to help.

We did not realize at that time what an undertaking it would be – and the rewrite moved erratically over the intervening years, especially after Gene's untimely death in 1994. Recently, however, we realized how important it was to complete it to help the growing number of playing card collectors and others with interest in the subject who needed the information.

In addition to those collectors mentioned by Gene as being of vast help in his original compilations, we have had tremendous assistance from a number of people and organizations. These include particularly:

- Shaffique Verjee to whom we owe a special thank you for allowing us to use so many of his scans.
- Rhonda and Bob Hawes for their diligent review of the material.
- The United States Playing Card Company; Ron Decker and Van Jones of the USPC Collection, and Margery Griffith retired curator of the Collection.

• Ellen Hochman	• Stuart Kaplan
• Esther Bollhagen	• Phil Bollhagen
• Jay Millman	• Lenny Schneir
• Bob Rosenberger	• Glenn Currie
• Toby Edwards	• Larry Herold
• Alan Ryan	• Jim and Dawn Faller
• Jack and Sharon Ferrell	• Larry Lubliner
• Gil and Barbara Adams	• Michael Goodall
• Ray and Pat Hartz	• Ron Greene
• John Lafler	• Dick and Sherry Sadler
• Doug Wuerth	• Wally and Tony Mach
• Ben and Jean Bornstein	• Steve Bowling
• Tony and Pat Fantilli	• Clara Huillade
• John McKinnon	• Neal Prescott
• Robert Simon	• Roger McCalmont
• Joe Friedman	• Dick Schuil
• Mike Shagan	• David Galt
• Phyllis Deane	• Jack and Mary Phillips
• Tom Ransom	• Rod Starling
• Bill and Betty Barnard	• Melissa McGroarty
• Pui Wah Cheung	• Paul Jung

Our deepest thanks to all of those mentioned and to everyone who contributed in any way to this Encyclopedia.

The title of this book is *The Hochman Encyclopedia of American Playing Cards* because it truly is Gene's work. The new items we have added that were unknown to Gene pale in significance when compared to the breadth of the whole volume. That we have been able to add many items, change the structure of the original work and publish it in one printed volume with improvements in quality over the original parts are modest achievements indeed when compared to the effort, research and dedication that went into Gene's original work.

We dedicate this book to Gene Hochman, a true pioneer in the dissemination of knowledge of this important facet of American life.

– Tom and Judy Dawson

COLLECTING PLAYING CARDS

Playing cards have been known since the mid-1300s, and it is a safe wager that almost since that time there have been people who were fascinated enough by the card images, artistry and folklore surrounding the cards to collect them. Certainly by the 17th century there were collections housed in museums and we can speculate that individual collections abounded as well. It is not the place of this Encyclopedia to describe the many fine collections that have been accumulated in museums and private hands over the centuries, but it is noteworthy that there now seem to be more collectors of old and unusual playing cards than ever before.

COLLECTORS

Collectors, whether casual or serious, fall into several categories. Firstly, there are those who collect complete decks of cards and those whose primary interest rests in collecting single examples of playing cards, whether for a court card, Joker or back interest. Then there are those who collect playing cards related to another main interest, for example Coca-Cola whose advertising decks are highly desired by Coke collectors, or gambling paraphernalia where American collectors like to add Faro or Steamboat decks to their displays.

Gene Hochman, in his earlier work had a few things to say about card collecting which are reproduced here.

"What about the new collector? Are there enough available decks for them to purchase reasonably enough to keep their interest alive without great expense? The answer is a definite YES!! A new collector of anything will start by trying to accumulate as many items as possible in their particular field. There are but two obstacles. One is the difficulty in finding collectible pieces and the more important of the two, the lack of available funds if and when they are located.

The term 'specializing' usually results when the collector has finally faced the second obstacle. A specialist must resist some offerings in order to save the available capital to purchase those gems that to him seem most desirable.

When I encountered my first 'Transformation' deck, I was so fascinated that I was determined to own it despite the fact that I knew I could not afford to. It was at that time that I sold several rare European decks to raise money to buy that first transformation deck. At that moment I became a 'specialist'. I have never regretted that decision and many years later, although I can now afford to widen my scope, I will still yield any of my rarities to secure an elusive missing item from any of my favorite categories."

These comments of Gene's still ring true although good decks are likely more expensive now on a relative basis. There certainly are more collectors and many of them are specializing, some by inclination and some by the very circumstance foreseen by Gene.

In his preface to Part IV, Standard Playing Cards, Gene wrote – *"There is nothing in this world like playing cards. What other games can be found, in one form or another, in over 80% of the homes in America, and a like amount in every country in the civilized world?*

Although basically meant for playing games, cards have been used for many other purposes. Their use as an advertising media has grown steadily. They have been used for Fortune Telling, by charlatans, for centuries. During World War II, they were used to instruct and help teach Aircraft recognition. Decks have been designed to aid the traveler in communicating in foreign languages. Legends are told of their use as a prayer book and an almanac. Just recently, I was at a gathering where a spontaneous raffle was necessary. The host called for a deck of cards, cut each card in half, distributed the top halves, and put the bottoms in a hat. A raffle book was made in seconds. I have always been amused by the tale of the explorer about to take a solo trip into unknown wilderness. He was asked why he packed a deck of cards. He answered 'if I ever get lost, I'll play a game of solitaire. There is sure to be someone looking over my shoulder to point out a move I miss.'

Collecting playing cards has been a rewarding hobby. I have enjoyed the excitement of the hunt for new decks. Every new table at every flea market may unfold the unexpected treasure. Most of all, I have enjoyed my fellow collectors, the most interesting people in the world. My trips into my private world of playing cards, miles and years from the cares of everyday living, is always at hand. This is what collecting playing cards has done for me.

I hope that this Encyclopedia will whet the appetite of the casual collector, and lead them down the path of pleasure which I have been travelling."

There is no question that Gene's Encyclopedia did whet the appetite of many collectors and make a huge contribution to the general improvement of this fascinating hobby.

There are a number of clubs supporting the playing card collecting hobby. Two of these are essentially American clubs, albeit with many overseas members; 52 Plus Joker, a club for deck collectors with an emphasis on American cards and The Chicago Playing Card Collectors' Club which caters to both deck and single card collectors. In Europe there are a number of clubs and England is the home of the International Playing Card Society, a group with members from around the world, whose main emphasis is on education and research into the history and use of playing cards.

CONDITION

Like any collectible, condition plays an important role in desirability and thus in value. We would all like our decks to be sparkling mint and still in their original wrappers and/or boxes. Unfortunately, most decks that collectors find have seen at least moderate use and have probably lost some element of their desirability.

While terminology relative to assessing the condition of playing cards has not been standardized, most collectors would agree that 'as issued' means the deck was found in about the same condition as when it left the factory. Perhaps it had been opened but never really taken from its packaging, and certainly never played with. If even the slightest element, e.g. a cellophane wrapper, is missing from an otherwise pristine deck, it could not be classified as 'as issued' – rather it would be 'mint'. If the missing element was of more consequence it would likely be further downgraded.

Gene Hochman devised a system to describe decks of playing cards:

- As issued – a complete deck, in mint condition, with all cards, jokers and extra cards contained in the original packaging when first distributed for sale. It might be unopened or carefully opened for examination, but not played with. If applicable, the tax stamp, not necessarily unbroken, would be attached.
- Mint – a complete deck showing no signs of use. Normally all cards would be present as would the original box in mint or near mint condition. The inside wrapper would not need to be there.
- Excellent – a complete deck that has been occasionally used, but still in first class condition. Gold edges would still be intact and you would be proud to use this deck in your game.
- Good – A complete deck showing signs of repeated use, but still useable. There would be no serious creases or bent/broken corners. The deck would not be swollen or misshapen and would fit comfortably into the original box.
- Poor – A deck not good enough to fit into one of the above categories. It likely would have at least one of these serious faults - bent or broken corners, bad creases, heavy soiling, etc
- With Faults – A deck in one of the good to as issued categories, but with a serious fault such as a missing or damaged card or a damaged, incomplete or missing box.

These descriptions were developed by Gene at the time of writing the original volumes and have stood the test of time. Many collectors have introduced variations into their cataloguing, e.g. 'mint plus', 'mint' and 'mint minus'. In addition, it has become popular to describe the condition of a deck's box as OB1 (basically mint), OB2 (some damage but complete) or OB3 (quite heavily damaged and/or some portion missing). Nonetheless,

use of the above descriptions and a careful notation of anything that is missing will provide an appropriate listing for cataloguing or selling purposes.

In all attempts to grade a deck, it is important to describe everything that is there and anything that is missing. For example, a brief description of an early advertising deck might read as follows:

"Advertising deck from 1910 for Dawson's Old Time Ale. Mint condition, in original box (slight damage to flap) with dated 2¢ U.S. revenue stamp. 53 cards with advertising Ace of Spades and special advertising Joker. The extra advertising card is missing and the Club Jack has a small smudge".

A note on missing cards. The extra cards over and above the regular 52 and joker(s) are clearly of less importance and a deck lacking one is hardly devalued, although the extra cards in wide advertising decks (which usually depict a factory, a separate ad, a price list, etc.) are more important. Again the pips in an important deck, especially one with unusual or non-standard courts, are of lesser importance than the courts. The Ace of Spades or Joker if missing, creates the most serious deficiency.

Despite most people's desire to collect only as issued, or perhaps mint, decks, collectors will still rejoice at finding a deck in only, say, good condition if it is high on their want list or quite scarce. Often it will be purchased with the expectation that the same deck in better condition will one day replace it

DATING OLD PLAYING CARDS

The following material, designed to assist collectors in dating their U.S. decks, is produced here courtesy of 52 Plus Joker. It appeared originally in an article by Margery Griffith, then Curator of the United States Playing Card Co. Museum in Cincinnati, in their quarterly bulletin *'Clear The Decks'* in April 1991.

"At first tax stamps on cards would seem an ideal way of pinpointing the date of manufacture of cards. This feature, however, has an area of inaccuracy ranging from zero up to 50 years or more, bearing in mind that old stocks of cards may not be released for many years; the blocks may be reused time and again, or even sold, the purchaser then making further packs."

*Sylvia Mann made this statement in her book, **All Cards on the Table: Standard Playing Cards of the World and Their History.** Ms. Mann spoke primarily about European cards, however the same holds true of their American descendants. For example, while re-examining our American cards in the process of cataloguing the USPC collection, a Samuel Hart deck was discovered dated 1868. It bore a 5 cent tax stamp (1872-83) covered over by a smaller 2 cent stamp (1894-1917). The cancellation overprint was "AD 1910" indicating that the deck was sold by Dougherty (not NYCC!) in 1910, showing a possible 40 year lapse between its manufacture and its distribution!*

While the dates given in the tax chart appearing below are generally accurate, additional information is necessary to use the chart effectively. In addition the cancellations on decks can help pinpoint the date of sale.

The tax rates, per deck, were as follows:

1862 to 1864	2 cents
1872 to 1883	5 cents
1883 to 1894	nil
1894 to 1917	2 cents
1917 to 1919	7 cents
1919 to 1924	8 cents
1924 to 1940	10 cents
1940 to 1941	11 cents
1941 to 1965	13 cents

In addition to the standard revenue stamps issued by the U. S. Government, there was another grouping of playing card revenue stamps that one rarely sees on a deck. These are the Private Die Playing Card Stamps issued from 1864 until 1883 when the revenue tax on playing cards was repealed. Under the Revenue Act of 1862, manufacturers were permitted, at their expense, to have dies engraved and plates made for their exclusive use. This method gave the manufacturers a slightly lower cost and the advertising value of the proprietary stamps could not be overlooked.

There are 16 different stamps in the Scott catalogue, numbers RU1 - RU16. They are listed below, along with their issue dates and total production, which indicates the size of the companies and number of decks made in that period:

RU1 - *Caterson Brotz & Co.* - 5 cents brown - first produced in 1882 but never issued - only 3 known
RU2 - *A. Dougherty* - 2 cents orange - May 1865 to July 1866 - 800,500 issued
RU3 - *A. Dougherty* - 4 cents black - December 1864 to September 1866 - 515,250 issued
RU4 - *A. Dougherty* - 5 cents blue (20x26mm) - August 1866 to 1877 - 12,450,428 issued
RU5 - *A. Dougherty* - 5 cents blue (18x23mm) - 1878 to 1883 - 7,980,983 issued
RU6 - *A. Dougherty* - 10 cents blue - December 1864 to May 1866 - 442,700 issued
RU7 - *Eagle Card Co.* - 5 cents black - 1880 to February 1883 - 1,800,900 issued
RU8 - *Chas. Goodall* - 5 cents black - November 1870 to August 1875 - 1,155,200 issued
RU9 - *Samuel Hart & Co.* - 5 cents black - September 1866 to 1877 - 8,129,053 issued
RU10 - *Lawrence & Cohen* - 2 cents blue - July 1865 to July 1866 - 1,149,750 issued
RU11 - *Lawrence & Cohen* - 5 cents green - July 1865 to March 1874 - 8,116,600 issued
RU12 - *John J. Levy* - 5 cents black - March 1867 to January 1873 - 3,124,840 issued
RU13 - *Victor E. Mauger & Petrie* - 5 cents blue -1877 to October 1880 - 1,021,020 issued
RU14 - *New York Consolidated* - 5 cents black - 1876 to March 1883 - 10,063,000 issued
RU15 - *Paper Fabrique Co.* - 5 cents black - June 1873 to October 1880 - 3,986,710 issued
RU16 - *Russell, Morgan & Co.* - 5 cents black - May 26, 1881 to March 22, 1883 - 1,304,100 issued

The different denominations result from the rates of tax during the period. While the tax was normally 2 cents to 1872, and 5 cents from 1872 to 1883, it is in reality more complicated than that. According to a series of articles written on these revenue stamps in 1931-32, the tax rates varied with the retail price of the playing cards from 1862 to 1866. The precise rates were as follows:

1862 to 1864

packs @ 18 cents or less - 1 cent
packs @ 19 to 25 cents - 2 cents
packs @ 26 to 36 cents - 3 cents
packs @ 36 cents or more - 5 cents

1864 to 1866

packs @ 18 cents or less - 2 cents
packs @ 19 to 25 cents - 4 cents
packs @ 26 to 50 cents - 10 cents
packs @ 50 cents to $1 - 15 cents

From 1866 to 1883 the rate was 5 cents a pack.

Keen readers will have noted that only Dougherty had a 10 cent stamp in the period 1864-66. The only other manufacturer with a private die stamp at that time was Lawrence & Cohen who would use two (or more) 5 cent stamps. In fact they had a 5 cent stamp when there was no 5 cent tax rate exigible! The other 5 cent stamps were all printed after the 5 cent rate came into existence in 1866.

Another dating aid, very useful for decks manufactured by United States Playing Card Co., was a dating code placed on the Ace of Spades at time of manufacture. This code was first published in Part VI of Hochman. The code first came into use in 1904 and it applies only to Aces of Spades that bear a letter plus a four-digit number. Combinations with fewer numbers have no meaning for collectors.

The letter code is as follows (updated from the original article):

A	1920	1940	1960	1980	2000
B	1921	1976	1996		
C	1922	1941	1961	1981	2001
D	1942	1962	1982		2002
E	1923	1943	1963	1983	2003
F	1924	1944	1964	1984	2004
G	1904	1925	1945	1965	1985
H	1905	1926	1946	1966	1986
J	1906	1927	1947	1967	1987
K	1907	1928	1948	1968	1988
L	1908	1929	1949	1969	1989
M	1909	1930	1950	1970	1990
N	1910				
P	1911	1931	1951	1971	1991
R	1912	1932	1952	1972	1992
S	1913	1933	1953	1973	1993
T	1914	1934	1954	1974	1994
U	1915	1935	1955	1975	1995
W	1916	1936	1956		
X	1917	1937	1957	1977	1997
Y	1918	1938	1958	1978	1998
Z	1919	1939	1959	1979	1999

Right from the beginning in 1904, the same codes were used by National Playing Card Co. and New York Consolidated Card Co., subsidiaries by then of USPC. Andrew Dougherty and Russell Playing Card Co. also used these codes, as they became part of USPC in 1907 and 1929 respectively.

Around 1965, USPC began the practice of "pre-facing" some decks, especially Congress decks. A supply of faces could be printed and stored and the backs could be added as needed. Therefore, Congress cards and any other pre-faced brands stopped using the codes altogether.

Decks were taxed based upon the number of cards per deck, jokers and advertising cards being exempt. One stamp was required for a deck of 52 or less cards; two stamps for decks from 53 to 104 cards (e.g. 64 card pinochle decks); double bridge sets required each deck to be wrapped and sealed with a stamp.

The cancellations on these stamps can be very useful to the collector, if not in determining the date, at least in identifying the maker. This is especially important for advertising or souvenir decks, or any deck which does not bear the maker's name but that of a publisher.

From 1940 to 1941 the tax rate was raised twice, from 10 to 13 cents. In order to disguise this increase, the government issued stamps saying "1 pack". Decks with the "1 pack" stamp can date anywhere from 1940 to 1965.

Finally on June 22, 1965, the tax on playing cards was revoked.

The use of tax stamps can be a very useful tool in dating a deck. Unfortunately, the collector often finds a deck where the tax stamp is missing or so defaced that it is illegible. In these cases the codes used by the USPC family of companies are helpful for decks produced between 1904 and 1965. Other sources of information include books, manufacturers and playing card collector club publications.

STORING AND DISPLAYING YOUR COLLECTION

It is important to take good care of your collection and handling and storage of your cards are key elements of proper care. When showing or looking at your decks be careful as you handle them, especially as you remove or replace them in their boxes and wrappers. In fact, many collectors store old wrappers in albums rather than risk damaging them as they look at their decks.

There are a number of stationery and archival stores where the packaging you need can be obtained. We can't emphasize too much the importance of using good archival materials (albums, wrappers, boxes, etc.) in storing your decks, especially your old and rare ones. After all, playing cards and their packaging are paper products and paper deteriorates with age. We owe it to ourselves, and those coming behind us, to do our absolute best to make sure our decks stay in the condition that we found them in for as long as possible.

There are a number of ways to store decks, including archival boxes, cases and spool cabinets or other chests with flat drawers. Decks that have no boxes should be packaged in some kind of protective cover, whether a plastic box, a paper wrapper, homemade box, etc. Elastics, unless of the new archival type, should never be used on a deck as they deteriorate with time and can cause considerable damage. Again, proper storage helps decks stay in their present condition longer and helps preserve them for the enjoyment of our future collectors.

KEEPING TRACK OF YOUR COLLECTION

Most collectors like to know what they have in their collections, so they record references to entries in playing cards books and keep track of values and other pertinent information about their cards. With the advent of the personal computer this has become a relatively easy task and most serious collectors now use some kind of computerized cataloguing system. Others use home developed cataloguing systems based on recipe cards, notebooks, etc.

It is normal to have at least the following information about your collection available for each deck:

- Manufacturer
- Brand name, including number, and name of deck, if applicable (e.g. Congress #606 - Moonfairy)
- Date of issue
- Box, wrapper, etc., whether or not there
- Number of cards and missing cards, if any
- Condition
- Type of deck (e.g. standard, advertising, souvenir, etc.)
- Cost and/or value
- Hochman or other reference
- Other pertinent data (e.g. gold edges, damaged cards, etc.)
- Catalogue reference number

EPHEMERA

Many playing card collectors also have a fascination for related collectibles, for example items associated with card games or gambling, items with playing card images or motifs, or the ephemera that is associated with playing cards. An attempt to list all known playing card related objects would fill several chapters and is beyond the scope of this book. Suffice to say, we have seen collectibles with playing card motifs in glass and porcelain, wood, paper, metals, fabrics, ivory, celluloid, plastics, and almost every other material known to man.

One area of interest to many collectors is commonly referred to as "playing card ephemera". The word ephemera likely means different things to different readers, but for our purpose we will define it as "items made from paper products which either use playing card images in a significant way or are associated with decks of playing cards or playing card manufacturers". We say paper products because ephemera, by definition, means something that was not made to last and that, in most instances, was expected to be discarded within a reasonably short time of its use.

Examples of ephemera using playing card images would include postcards, general advertising, fruit labels, trade cards, calendars, posters and store cards. Ephemera more directly related to playing card manufacturers and their products would include product advertising, wrappers, extra cards, letterheads and envelopes, billheads, packaging and display boxes, booklets, trade cards, catalogues and price lists, salesman's samples, and playing card tax stamps. Several

items of playing card ephemera are scattered throughout the Encyclopedia and the colored plates.

Like collecting the cards themselves, collecting playing card ephemera provides almost endless variety. This special area has become, for many card collectors like us, one of significant interest and fascination. It provides one with much more insight into the makers, and the quality and beauty of the ephemera itself adds greatly to the collector's interest. It is interesting that, amongst people who collect various types of paper ephemera, items printed by the playing card manufacturers, especially those from about 1870 to 1920, are in great demand because of the high quality of the printing and lithography used by these quality printers in promoting their own products.

Playing card ephemera was made to be discarded after its intended use was over. Most of it is even more fragile, and consequently needs more tender care and preservation, than playing cards themselves. Storage in appropriate archival materials is a must, if one is to preserve the item to the best of one's ability, thus preserving it for others to enjoy in the future and, incidentally, preserving its value.

VALUING OLD PLAYING CARDS

Gene Hochman, in Volume I of his Encyclopedia, included a price guide for all the decks listed. Oh, that we could purchase these for the prices he quoted at the time! We have prepared, to complement this Encyclopedia, a separately bound price guide in which we use three of the categories discussed earlier to describe the listed decks. These are:

- Mint – a complete deck showing no signs of use. Normally all cards would be present as would the original box in mint or near mint condition. The inside wrapper would not need to be there.

- Excellent – a complete deck that has been occasionally used, but still in first class condition. Gold edges would still be intact and you would be proud to use this deck in your game.

- Good – A complete deck showing signs of repeated use, but still useable. There would be no serious creases or bent/broken corners. The deck would not be swollen or misshapen and would fit comfortably into the original box.

Prices for decks in the other categories can be interpolated from those shown. For example, a deck that is 'as issued' would command a premium over the mint price. Conversely a deck that is poor would be worth less than a 'good' one, and one with faults would likely be subject to a significant discount.

There are still quite a number of decks where the number of known copies can be counted on the fingers of one hand. Many of these are in museum collections and many of the very early decks, listed 'mint' may not even exist in that condition, but the category is priced on the basis that one or more may become available in the future.

When using this guide in determining the value of any deck of cards, keep in mind that, while it has been compiled from auction lists and decks offered for sale by antique dealers, internet and other auctions, rare book shops and private collectors, prices are nonetheless somewhat subjective. As sales of rarer decks are few and far between, a particular collector's desire for a certain deck can often result in an unrealistic price. Or, the sudden entry on the market of a few copies of a scarce deck can result in sales at prices substantially less than previously obtained.

We have tried to take note of decks that appear to be present in most collections and those that are scarce and wanted by many different collectors. Prices must also be based on the number of collecting fields an individual deck might encompass. For example, a baseball deck would appeal to baseball nostalgia collectors as well as playing card collectors. An advertising deck from the Columbian Exposition might be sought by World's Fair and advertising collectors as well as those in our field. In the final analysis, scarcity of the item, the law of supply and demand and condition will determine the price.

In Gene's last issued price list in 1991 he presented some advice for both buyers and sellers. It was, and still is good advice, and we repeat it here: *"advice to buyers .... if you see a deck that you really want for your collection and you have an opportunity to buy it, and the price seems higher than the listed value, remember you may never find another .... and if you do, it will probably be for more. Even if you overpay slightly, it will not be long before the value will surpass the purchase price. Advice to sellers .... using this list as 'the price' you must get, will result in many lost sales. You must find a collector looking for a particular deck and willing to pay your price. It may pay to wait, but if you must sell quickly, be prepared to take less."*

The demand for old and rare playing cards far exceeds the supply, and we have all experienced regret, on occasion, for not paying the additional dollars necessary to purchase a scarce deck that we have not had another chance to buy.

All prices are based on complete decks, with Jokers if so issued, and in the original boxes, if sold boxed. Any faults or defects, of course, reduce the value and decks in mint or as issued condition will almost always bring a premium.

An important fact to remember when using the price guide, is that it is only a guide. It is also only our opinion, which is nonetheless based on years of research, constant study of prices realized at auctions, and our general experience gained in over 25 years of collecting. We also prevailed upon a number of collectors to add their knowledge and expertise to the process in their areas of specialization. The result, we believe, is a reliable and fair price guide.

The prices of standard decks are more difficult to estimate than those established for popular categories such as transformation and insert cards, souvenir and railway cards and advertising decks. Non-standard decks usually have beautiful and/or interesting courts, can be of historical significance and often appeal to more than one group of collectors.

The law of supply and demand, therefore makes standard cards the cheapest available. Many of the decks listed in this book in Chapters 3 to 16 are rarer (several are likely unique) than decks selling for much higher prices in other categories. Novelty, souvenir, political, transformation and other non-standard decks were often preserved and placed in a hidden nook to be found and collected many years later. This was not usually the case with standard decks. They were invariably bought to be used, and were almost always discarded when soiled or worn.

Only ardent collectors seem to have any interest in the varieties of standard decks, a great mistake in our opinion, as they are the backbone of the history of playing cards. This impacts the pricing of standard decks and readers will note that a rare 19th century standard deck may have a price, for example, less than that of a 'war' deck with many, many more copies of the latter in the marketplace.

A final point. We believe, like Gene did, that values for good decks will only rise. Scarce items only become scarcer, and as more people realize the joy of collecting playing cards and become serious collectors, the demand for old and scarce decks, especially ones in excellent or better condition, will continue to grow and drive prices upward.

ORGANIZATION OF THE ENCYCLOPEDIA

The original Encyclopedia of American Playing Cards was published in six volumes between 1976 and 1981. It was Gene Hochman's dream to re-publish the Encyclopedia, with many additions and corrections, in one large comprehensive book. He wanted the book to be useful to, and used by, not only collectors of United States playing cards, but also all people with an interest in the subject. He therefore wanted it to be "user friendly" to collectors and of sufficient import that it could, and would, be used by those studying or researching American playing cards.

These goals are, to a certain extent, mutually exclusive. Too much detail and information would make the book bulky and more difficult for the typical collector to use. On the other hand a book geared just to the identification of specific decks would be of little interest to the student of American playing cards and their makers.

We have tried to meet Gene's wishes and accommodate both groups in producing this volume. All known decks by American manufacturers are listed and pictured, although in cases of a few rarer decks, we might not have been able to obtain the access required to photograph the cards. For the more serious student of American cards, we have done additional research into the known manufacturers and obtained much more information than was available to Gene at the time of the original writing, most of which is included in this book.

For standard decks we have usually pictured the Ace of Spades and the Joker. Where there are other cards we judged to be of significant interest, we have shown them as well. We have maintained Gene's concept of using the King of Hearts for identification purposes in identifying decks from the early makers and other mystery manufacturers of the 19th and early 20th centuries.

There are some decks where we have been able to obtain a photograph of the Ace of Spades but other important cards, e.g. a Joker, have not been shown because they were not available to us. While every attempt has been made to provide high quality pictures through new photographs, this has not always been possible. We have used a variety of sources and methods to obtain what you see - scans of actual cards being the most common, although we have, in several instances used scans of photostats or photographs and even scans of pictures in the original volumes.

An important decision was the extent of coverage of 'newer' decks. After consultation with Gene, we decided that we would, as a general rule, restrict coverage of non-standard decks to the period from the origins of manufacture in the United States, about 1800, to the 1930s. The exceptions would be mostly those falling into the WWII and later period, which he had already listed in his original volumes. In the case of standard brands produced in the 1930s, or earlier, we would attempt to continue describing their evolution up to the present date. Exceptions to, and clarifications of, these general guidelines, are described in the chapters pertaining to particular categories.

The reader will recognize that the Encyclopedia is still incomplete. We have no doubt that previously unknown variations, and indeed new brands and non-standard decks, will continue to be discovered. That is part of the excitement of this collectible. We are also sure that additional information will come to light that will help better identify certain decks and makers. That is desirable as well, as we share Gene's thought expressed in Volume I - *"No doubt I will have learned of omitted decks before this edition reaches the press. But, the foundation will have been laid. The readers and I will be able to build and improve our knowledge with greater ease with something available to use as a guide."*

Those were prophetic words in 1976. There have been countless new discoveries but collectors have been very well served by Gene's work. We hope this volume will continue to build on Gene's foundation.

The book has been organized as follows:

- Chapters 3 to 15 cover standard decks with the earliest manufacturers in the earlier chapters,

- Chapter 16 is new material containing a brief history of Canadian manufacturers and listing of known Canadian decks,

- Chapter 17 deals extensively with advertising decks,

- Chapters 18 and 19 deal with the old Volume II categories of transformation and insert decks,

- War and political categories are covered in Chapters 20 and 21,

- Chapter 22 covers the entertainment category,

- Chapters 24 to 27 deal with exposition, souvenir and railway decks,

- Chapters 23 and 28 to 33 cover the remaining categories of the old Volume III - bridge/whist, no-revoke, colleges and unions, fortune telling and tarot, oddities, etc.

Finally, we have included a comprehensive index of decks and manufacturers.

We debated long and hard whether to change the reference system and entry references used in the original works. Gene's reference system is quite workable and does not need amendment except in one major area, and that is ordering. His original concept of listing standard decks by maker, in date order, does not work effectively, given the large number of new discoveries and additions since Volume IV was published. However, for a variety of reasons, particularly the extensive use of the existing references in books, catalogs and collectors' databases, we decided to continue the present system.

Nonetheless, we had to make some changes to fit the circumstances. For example, many of the decks previously listed in the Miscellaneous Standard Wide (MSW) category have been placed in their own Chapter (e.g. American Playing Card Company MSW1 to MSW18 now part of the Longley group in Chapter 4). Other

MSW decks have been found to be made by major makers. In cases where we have changed the original reference number we have included a reference to the former listing, e.g. "**NY81 LIGHTHOUSE #922** (formerly MSW95), c1920". We have also included a table in Chapter 13, cross-referencing the old reference to the new one.

Finally, we have added new listings, as subsidiary to existing listings, where we judged this was the clearest way to proceed. This would be true, for example, if we found a 'new' L. I. Cohen deck, where it would be more sensible to list it as NY7a, rather than adding it to the end of the NY listings as NY88.

THE EARLY MAKERS

We have all read stories and legends about the first playing cards in America. Cards made of leaves were used by seamen of the Columbus Expedition, after throwing theirs overboard in superstitious haste. Card collectors have seen or read about Indian playing cards made of animal skin. A good number have been found among the Apache Indians, other U.S. tribes and, more recently among South American Indian artifacts. These are far too crude to be considered manufactured.

There must have been many lonely hours at the outposts of civilization. Card playing was surely one of the main recreations. Early printed cards were brought, in large quantities, to settlements in the New World. English cards were probably used in Virginia and New England. French cards found their way to Canada, Dutch cards to New Amsterdam and Spanish cards to Florida, Mexico and the Caribbean. Obviously one cannot tell, should one of these early imports be found, if it was indeed used in America, as these decks would not be labeled 'used in America.' The Aces of Spades of early English decks made for export were marked 'EXPORTATION', but never to where. As the Encyclopedia covers playing cards manufactured in America, only imports having a direct influence on American manufacture will be mentioned.

An overwhelming percentage of early American Colonists were British subjects. They eagerly awaited the arrival of English manufactured cards in order to indulge in their favorite pastime. Cargo ships arriving from England always listed a goodly amount of playing cards. Until recently, a nice reproduction of a deck manufactured in England for export to America by I. Hardy could be bought at the restoration site in Williamsburg, Virginia. It is easy to distinguish the copy from the original, but the details are quite accurate. The deck is packaged in a copy of the original wrapper, including the hated tax stamp.

When were the first playing cards manufactured on this side of the Atlantic? There is no exact answer to this question. Each fact uncovered raised a new question. Let us consider some facts. Firstly, before the Revolution, no more than 50 printing presses were in the Colonies. Secondly, the paper needed to produce playing cards was not available and no cargo manifests listed special playing card paper. Thirdly, newspapers, posters, signs, business forms, currency and many other important items would surely have been given preference. Finally, there were ample amounts of imported cards, of superior quality, available at a reasonable price.

Based on a variety of research, it can be safely concluded that the first American made playing cards were not offered for sale until after 1776. In the interest of accuracy, we offer some Questions (Q), Speculations (S), and Facts (F), that summarize the conclusions arrived at in the research.....(F) After the signing of the Declaration of Independence, there was little or no market for playing cards bearing the hated English tax stamps or having English tax Aces of Spades. (F) There was a large and ever increasing demand for playing cards in America. (Q) How was this large demand for cards filled? (Q) What was done with the large stocks of English made playing cards on hand at that time? (F) Many early American decks have court cards closely resembling those of English packs of that period. (S) Aces of Spades were printed here, along with new wrappers and as there were no 'Truth in Advertising' laws, the newly made items were used to replace the original tax Aces of Spades and wrappers. The decks were then offered to the public for sale as 'American Manufacture'.

It is safe to assume that the decks found today with nondescript Aces of Spades, which simply say 'American Manufacture', were made in America. In an effort to determine the earliest makers, business directories of the large Eastern cities, newspaper advertisements, and all of the large collections of U.S. cards were examined. It should be noted, that any printer having a press, any stationery store making business cards, and countless others, could have tried to become a playing card manufacturer either directly, or by commissioning decks from other printers or playing card makers.

After gathering much valuable information there are still a number of early makers about which little is known, with no cards identified as examples of their craft. In addition, while some 'American Manufacture' decks can be reliably attributed to specific makers, others cannot, for example many of those dedicated in the original Encyclopedia as a 'Tomb of the Unknown Card Maker'. Decks that have been identified since then reappear in other chapters, but some remain a mystery. Finally, many more decks from early makers have been discovered.

All of this additional information, while helpful, creates more confusion in many respects. In an effort to help establish order out of the confusion, we have listed the known early makers below - by location, with reference numbers to decks in this chapter where applicable. Incidentally, early makers, for purposes of this chapter, include 19th century decks, normally without indices, that are not attributed to either the Longley, Andrew Dougherty or Cohen/NYCC groups covered in Chapters 4, 5 and 6.

Area	Maker	Dates	References
Boston	Amos Whitney	1795 - 1811	U1b
	William Coolidge & Co.	1811	U1c
	Jazaniah Ford	1793 - 1839	U1d
	Joseph Ford	1811 - 1814	U1e
	Nathaniel Ford	1839 - 1847	U1, U1a
	Thomas Crehore	1802 - 1846	U2, U3, U4, U5
	Samuel Avery	1840	
	Chas. E Townsend	1860	
Philadelphia	Edward Ryves	1776	
	James Robertson	1778	
	Lester & Poyntell	1779	
	Ryves & Ashmead	1790	
	J. Rice	1796	
	Thomas de Silver	1811	
	C. Buffard	1816	
	K. Fizel	1816	
	J. Y. Humphrey	1816 - 1830	U29, U29a, b, c & d
	Saml. M. Stewart	1840	U31
	David Gilbert	1845	
	Charles Bartlet	1845	U14
	Continental Card Co.	1874 - 1880	U18, U18a, b & c
New York	John Casenave	1801 - 1807	U33
	Henry Hart	1800 - 1825	U41
	David Felt	1826	U6
	Robert Sauzade	1830	U32
	Caleb Bartlett	1830 - 1850	U7, U8, U11, U12, U12a, U13
	N. Y. Card Manufactory	1830	U9, U10
	Boston Card Factory	1830	U9a
	Congress Card Manufactory	1852	U15
	Strauss & Trier	1860	U27
	Sterling Card Co.	1876	U37
	J. Thoubboron	1870	U17
	Regenstein & Roesling	1870	U26
	Victor E. Mauger	1873 - 1878	U19, U19a, b & c
Bergen County, N.J.	D. C. Baldwin	1789	
Unknown	Henry Hart	1810	U39
	Hunt American Manufacture	1812	U38
	J. & L. D.	1840	U34
	Carmichael, Jewitt & Wales	1840	U35, U36
	Jones & Company	1860	U30

The balance of this chapter discusses the known makers and depicts those decks that have been attributed to them. Wherever possible, the King of Hearts will be used from the illustrated decks for comparison purposes.

The early Boston makers consisted of Amos Whitney, the Foord Family (who spelled their name on their cards with one 'o') and Thomas Crehore. Jazaniah Ford has been credited by many as being the first American playing card manufacturer, but this is not correct.

Professor and Mrs. Wayland of Pasadena, California, conducted extensive research into the early New England manufacturers and most of the information and dates in the table above are taken from their work. It appears therefore that Ford was the first in the Boston area, but that he started several years after the earliest makers in Philadelphia.

The Wayland's obtained vast amounts of information about the Ford Family. Using advertisements, obituaries, cashbooks, business directories, invoices, etc, it was found that Jazaniah Ford manufactured playing cards in Milton, Mass. from 1793 until his death in 1832. His son, Jazaniah Seth Ford carried on until his death in 1839 and then a cousin of Jazaniah Seth, Nathaniel Ford kept the factory going until 1847.

Records of the company and copies of their advertisements state that many types of decks were made. In the original Encyclopedia, the only Jazaniah Ford decks attributable to that maker were the Decatur deck of 1812 and the Lafayette deck of 1824 (refer Chapter 20). We are now aware of a few standard Jazaniah Ford decks in the hands of collectors and we know that they issued several brands such as Lafayette Eagles, Eagles, Highlanders, Merry Andrews and Henry VIII.

Joseph Ford, Jazaniah's brother, was found to have manufactured playing cards in his own plant, in Milton, Mass., from 1811 until his death in 1814. Books of the company show that he was heavily in debt, and apparently unsuccessful. Only a single Ace of Spades, in the collection of the Society for the Preservation of New England Antiquities, remains. Other than a change of name, the Ace is identical to those of Nathaniel Ford.

It is likely more than a coincidence that the start of Joseph Ford's business followed immediately the demise of the playing card business of Amos Whitney. Whitney started making cards in 1795 when he was 32 and continued making such grades of standard cards as Superfine Columbian, Harry the Eighth and Merry Andrew until he died in 1804. The business was then run by his father Daniel Whitney until his death in 1810, and then briefly by William Coolidge. In September 1811, the business was auctioned and, quite likely, was the foundation on which Joseph Ford tried to build a successful manufactory.

Nathaniel Ford, as we saw, made playing cards for about eight years from 1839 to 1847. A few examples of standard cards with his Ace of Spades, both normal and patience size, are known. However, no cards can be found that can definitely be identified as having been from a deck with the name Jazaniah Seth Foord on the Ace, although a wrapper from a deck bearing his name exists (the wrapper shows that he inserted the second 'o' back into the family name).

The only evidence of Whitney cards, is an Ace of Spades at the Society for the Preservation of New England Antiquities and one in the Cary Collection at Yale University accompanied by several court cards that appear to be from the same deck. We do know Whitney issued several brands including Columbian, Harry the Eighth and Merry Andrew. The Cary Collection also includes a full deck with an identical Ace of Spades, except for the name, made by William Coolidge at the same Orange St. address used by the Whitney's. It is safe to assume that Mr. Coolidge took over the Whitney business after Daniel's death until the auction in 1811.

We have decided to put all the early Boston area decks under the U1 listing to help preserve the original references and keep things as close to chronological order as we can.

U1 NATHANIEL FORD & CO.
Milton, Mass., c1840. A standard deck for the time with square corners and one way courts.

U1

U1a

U1a NATHANIEL FORD
Milton, Mass., c1840. A different Ace of Spades on a miniature deck measuring 2 1/16 inches by 1 inch. The courts are quite different than U1.

U1b

U1b AMOS WHITNEY
Boston, c1798. The Ace of Spades and three courts are in the Cary Collection in New Haven.

U1c

U1c WILLIAM COOLIDGE
Boston, c1810. Another deck in the Cary collection with essentially the same Ace as U1b.

U1d

U1d JAZANIAH FORD
Milton, Mass., c1830. A standard deck by this maker.

U1e

U1e JOSEPH FORD
Milton, Mass., c1811. No complete decks are known but the Ace of Spades is very similar to the ones used later by Nathaniel Ford.

Thomas Crehore was the first of the 'patriarchs' to have left behind considerable evidence of his cards. He began, c1798, in Dorchester, Mass. and when he retired his sons continued his business until the factory burned down in 1846. Crehore invented new methods of playing card manufacture which were apparently successful, judging by the many decks that still exist. There are at least four different Aces of Spades carrying the Crehore name, including one bearing a misspelled version, now known to be attributable to a later maker trading on the Crehore name.

As can be seen, the incorrect spelling of his name was the omission of the final 'e'. There are also reports of Aces bearing the names Isaac Crehore and Edward Crehore, the sons who carried on the family business after Thomas' retirement. We have been unable to find one of these decks, or any collector who has seen one.

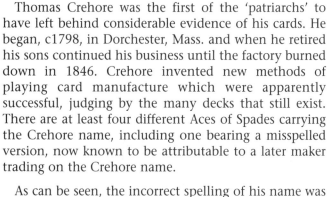

U2

U2 THOMAS CREHORE
Dorchester, Mass., c1820. This Ace of Spades and the two that follow have slight differences but we cannot be sure which one was issued first. Both U3 and U4 have only 13 stars on the Ace compared to 16 on U2, and while some believe "the more stars the later the deck", other authorities do not subscribe to this theory. These three decks all have similar, single-ended courts, the base for the common Pharo (Faro) courts popularized by Samuel Hart and others.

U3

U3 THOMAS CREHORE
Dorchester, Mass., c1820.

U4

U4 THOMAS CREHORE
Dorchester, Mass., c1820.

U5

U5 THOMAS CREHOR
c1845. These decks can be found with two quite different sets of courts. It is likely that the more common version (like U4) was manufactured by Samuel Hart and the above by Andrew Dougherty.

U5a

U5a AMERICAN MANUFACTURE
c1845. This deck is so similar to one of the Crehor decks that we have decided to list it here.

Amongst the brands produced by Crehore over the years were Columbian Eagle, Harry VIIIth, Merry Andrew, Highlander and Black Eagle.

In the 1840s, cards with remarkably similar court cards were made with different Aces of Spades purportedly by Thomas Crehor, Andrew Dougherty and Samuel Hart. In addition, Charles Bartlet and Hart made similar decks. Research has recently demonstrated that Hart and Bartlet were one and the same and that the Crehor Aces were placed on decks made by both Dougherty and Hart. The following information summarizes the basis for these conclusions and provides additional insight into the enigmas surrounding the cards manufactured in the 1840's under all four of these names.

- Andrew Dougherty manufactured decks with courts identical to those found on some Crehor decks. In addition, comparison of minute details on Dougherty and Crehor pip cards has found identical flaws.
- Hart left the employ of L. I. Cohen, his uncle, in 1844 to start his own business.
- The Bartlet courts (U14) are nearly identical to the Hart courts listed as NY24 (minor changes are a natural evolution as plates are redone frequently).
- Bartlet's address in Philadelphia was 236 S. 13th St. - the identical address to Hart's in 1846-48.
- The Charles Bartlet cards were made by Samuel Hart in an effort to trade on the reputation of Caleb Bartlett, who had just stopped producing. Hart also used the name Thomas Crehor having purchased machinery, plates, etc. from Crehore after fire destroyed the company.
- As Hart seemingly was "trading on" the old Crehore name, it is not surprising that he would, in the initial years, try to "trade on" the Caleb Bartlett name as well.

We have retained the listing for Crehor in this chapter and have also referenced it in the Dougherty and NYCC chapters.

Early business directories of New York City, from the late 18th to early 19th centuries have been examined in an effort to determine who was making or selling playing cards. There were several competing companies making these directories and some names, found in one, were missing from another making one wonder if there were small entrepreneurs who were not listed in any. Advertisements in early newspapers and periodicals often mentioned playing cards under listings of printers or stationers. Even if described as makers of playing cards, it is possible that others manufactured for them, using specially designed wrappers and Aces of Spades.

One conclusion from this is that it is virtually impossible to precisely date any deck from the first half of the 19th century. Where there are several specimens, sometimes by comparison we can determine which came first, but as we saw with the Crehore decks above more often than not, we cannot. Dating, of course, became much easier with the arrival of tax stamps in 1862. Dating became quite accurate when the decks were found in their original cases or wrappers, although decks delivered and stamped many years after manufacture are known.

David Felt, a New York City stationer, advertised playing cards, of his own make, in 1826. A full deck, in the original wrapper, is in the collection of the United States Playing Card Co. Pictured are the Ace of Spades and the King of Hearts. It is odd that the wrapper is nearly identical to the one used by Jazaniah Ford for his Decatur cards and that it mentions 'Decatur Cards' and 'Harry the VIII' and 'Eagles'...all brands which were sold by the Ford Company and listed in the ledgers of Jazaniah Seth Foord. Perhaps Felt really was not a maker and had these cards made for him by the Ford factory.

U7

U7 CALEB BARTLETT
New York c1830. Through comparison with many Caleb Bartlett decks, this seems to be the first one he produced. Note the "CB" initials on the King of Hearts - the C is just below his left arm and the B is in the lower right portion of the same section. The other courts also have the initials, many cleverly hidden.

U6

U6 DAVID FELT
New York, 1826. The name did not appear on the Ace of Spades, only the wrapper.

U8

U8 AMERICAN MANUFACTURE
(Caleb Bartlett), 102 John St., New York, c1830. The identical court cards as U7. There is a striking resemblance in the basic design to those of Thomas Crehore who manufactured playing cards until 1846. Caleb Bartlett ceased manufacturing his cards in 1842.

Caleb Bartlett was one of the most prolific producers of playing cards of this early period. Tracing his exact street locations, by means of early NYC directories, was like following a bouncing ball. There were locations on Pearl St., Bowery, Fulton St., Broome St., Broadway, 4th St., Staten Island and John St. to name several. Optimists can assume that it was due to his consistent growth.

Caleb Bartlett's standard packs are unusual because the initials 'CB' are discretely placed in some location on each court card. The initials have allowed us to establish that New York Card Manufactory and Caleb Bartlett were one and the same. They have also positively identified some of the decks issued with the plain 'American Manufacture' Ace of Spades. The Bartlett courts were continued in use by Charles Bartlet, Thomas Crehor and Samuel Hart and are often referred to as Hart Faro courts.

U9

U9 N.Y. CARD MANUFACTORY
(Caleb Bartlett), 102 John St., New York, c1830. The same John St. Ace of Spades was used, but with the addition of the words "N.Y. Card Manufactory" (see U10). The use of the initialed courts has enabled us to state positively that Caleb Bartlett and the N.Y. Card Manufactory were one and the same.

U9a

U9a BOSTON CARD FACTORY
(Caleb Bartlett), New York, c1830. An almost identical Ace with the same initialed courts, allowing us to positively identify the maker.

U10

U13

U13 CALEB BARTLETT
Abbot & Ely, c1842. Ely seems to be the link with the U12 deck and perhaps this deck is earlier. Abbot & Ely were best known for their reprint (W2a) of the Indian Wars deck by Humphreys. The Abbot & Ely deck was the one reproduced by Sturbridge Village for the Bicentennial celebrations (NR3).

U10 N.Y. CARD MANUFACTORY
(Caleb Bartlett), c1840. This deck was issued with French style courts. As settlers arrived in the New World, there were many efforts to offer products, including playing cards, that would appeal to the newly arrived consumers. The Ace of Spades was the same as U9, but without the John St. address.

U11 CALEB BARTLETT *(not pictured)*
c1835. One of the most interesting decks produced by Bartlett was his transformation pack of 1833 (see Chapter 18, Transformation Cards). The plates used in the Bartlett Transformation deck (T1) were very expensive to make although based on the identical patterns first used by the Ackermann Repository of Arts, in London, in 1819 and later used by Muller of Vienna and Gide of Paris. Perhaps it was for this reason that Bartlett produced this quasi-standard deck using the transformation courts and aces with standard European style pips.

U14

U14 CHARLES BARTLET
Philadelphia, c1845. As concluded earlier, this maker was really Samuel Hart and is referenced in Chapter 5 as well. The Ace of Spades is always the same, but the courts vary from quite elaborate (shown) to the Hart Faro type.

U12

U12 ELY, SMITH & COOK
late Caleb Bartlett, c1843. This deck used the same, initialed courts as U7 to U9 and we might assume they bought the company from the estate and continued the Bartlett and NY Card Manufactory names.

U15

U15 CONGRESS CARD MANUFACTORY
New York, 1852. The pictured deck was produced for C.R. Hewet, a stationery store in New York City. There is very fine detail on the Ace of Spades. A similar deck, labeled Congress Manufacturing Co., Astor House, N.Y. is in The US Playing Card Co. collection. This company was listed in the New York City directories from 1853 to 1855.

U12a

U15 WRAPPER

U12a GEO. COOK LATE WITH ABBOT & ELY
c1845. Another deck with the same Bartlett courts and also using the Ely and Cook names, but likely produced a bit later.

U15a CONGRESS CO.

c1850. This deck is rather poor quality and has the Hart Faro type courts. While it might have been manufactured by the Congress Card Manufactory, it might just as likely have been made by Dougherty or Hart. Quite possibly it is a marked deck as the back is of a type often marked in that era for card sharping. Naturally, reputable manufacturers were loath to use their name on marked decks and they are often found with innocuous ones like this.

U15a

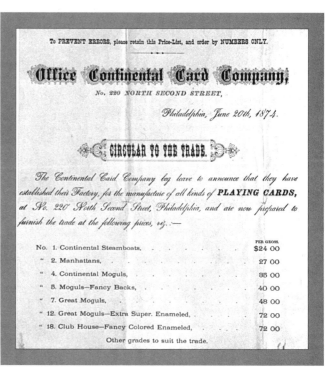

U18 WRAPPER

The Coughtry & Dougherty listing, U16, has been determined to be a venture of Andrew Dougherty and has therefore been moved to Chapter 6.

U17

U17 J. THOUBBORON

28 Cedar St., New York, c1870. The pack shown here is the only one known by this maker and no trace of this company has been found. The courts and Ace of Spades are somewhat similar to those used in decks made by Andrew Dougherty and Lawrence & Cohen. Neither one, though, is close enough to infer that one of them made the cards for Thoubboron.

In any event, there is a copy of circular to the trade dated June 20, 1874, announcing the formation of the company "for the manufacture of all kinds of playing cards at No. 220 North Second Street, Philadelphia". Brands advertised included:

No. 1 Continental Steamboats

No. 2 Manhattans

No. 4 Continental Moguls

No. 5 Moguls – Fancy Backs

No. 7 Great Moguls

No.12 Great Moguls – Extra Super, Enameled

No.18 Club House – Fancy Colored, Enameled

U18

U18 CONTINENTAL CARD CO.

Philadelphia, 1875. This deck of cards, made for Faro, was found in its original wrapper and the unbroken tax stamp was dated 1894, but we can assume that the deck was made earlier. The King of Hearts and the rest of the courts are very similar to decks produced by the NY Consolidated Card Co. (Samuel Hart) at the same time, and Continental was perhaps a firm name created by Samuel Hart for use in Philadelphia after the combination into NYCC.

CONTINENTAL TRADE CIRCULAR

U18a

U18b

U18a CONTINENTAL CARD CO.
Philadelphia, 1876. This deck, made for the Centennial, is also listed in Expositions as SX3.

U18b CONTINENTAL CARD CO.
Philadelphia, c1874. Likely the earliest deck made by this company.

U18c

U18c CONTINENTAL CARD CO.
Philadelphia, c1875. This deck was found with a "Highest Trump" Joker.

As there are very few decks around by this maker, we can assume that they did not stay in business very long.

Our next manufacturer, Victor E. Mauger was born in England and emigrated to the United States in 1855 to set up business as an importer of metal goods. In his leisure time he supplemented his income by writing articles for the press, bringing him into contact with the printing and stationery trades. Shortly thereafter he started importing equipment for these fields which led to his involvement with playing cards. Initially, Goodall's cards, which were imported and sold by Mauger commencing in 1867, were standard patterns from the Goodall range. By the early 1870s Goodall manufactured a special Euchre pack exclusively for America with a Goodall Ace of Spades and a multicoloured seated Joker with the inscription C. Goodall & Son, London and New York (U19c). By 1873 this enterprising man was manufacturing his own cards to complement the Goodall line (which was dropped in 1876 due to high import duties). Mauger continued making cards, in latter years under the name Victor E. Mauger & Petrie, until he sold his business to rivals Andrew Dougherty and NYCC in 1878. Ironically, in 1880 he was hired to set up the playing card department for Russell, Morgan & Co. and worked with it, and its successors, as General Agent into the 1890s. He finished his career back in his native England as the United Kingdom representative for USPC where he ran The American Playing Card Agency.

U19

U19 VICTOR E. MAUGER
New York, c1873. Mauger is famous for his special four color Quadruplicate pack issued in 1876 for the Centennial (see SX1). His standard Ace of Spades also carried the name of Goodall, London. There are at least three variants of the colorful Joker.

U19a VICTOR E. MAUGER
A deck of Mauger cards, found in the original wrapper, which has the identical Ace of Spades as U19. The courts are one way and quite unusual, as they have an excessive amount of gold color. It appears they were attempting to create an 'illuminated' deck, but used a yellow printing instead of gold leaf.

U19a

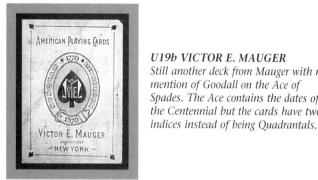

U19b VICTOR E. MAUGER
Still another deck from Mauger with no mention of Goodall on the Ace of Spades. The Ace contains the dates of the Centennial but the cards have two indices instead of being Quadrantals.

U19b

U19c

U19c COLUMBIAS
Chas. Goodall (for Victor E. Mauger), c1872. This Euchre deck was made expressly for sale by Mauger in the United States and in fact the wrapper mentions that Mauger had applied for a U. S. patent for the deck.

Special brands imported by Mauger for Goodall for the United States included Steamships, Regattas, Broadways, Virginias, General Jacksons, Jacksons Sporting, Columbias, Golden Gates and Mount Vernons. Other brands produced in the United States included Steamboats Quadrantals, Grand Pachas, Quadrantal Highlanders, Quadrantal Moguls, Quadraplet Maugers and Chaumette Pointers.

Finally, it is interesting to note that Victor Mauger advertised as a commission merchant at 82 St. Peter Street in Montreal, Canada in 1870. The ad indicated it was a branch of 110 Reade Street, New York.

EARLY MAUGER WRAPPER

The Longley Brothers, Paper Fabrique decks listed as U20 to U25 are now listed in Chapter 4.

U26

U26 REGENSTEIN & ROESLING
537 Pearl St., New York, c1870. This deck of French Piquet cards is in the collection of USPC. The Ace of Spades is a plain spade pip and the courts are similar to German standard. The wrapper with the name was found with the pack but no mention of the manufacturer's name appears anywhere on the cards themselves. As original wrappers are rarely preserved with used decks, it would be logical to assume that many decks of similar origin, are marked "anonymous" and are now in drawers labeled France or Germany. Check your collection!

U26a REGENSTEIN & SCWARTZ
New York, c1875. Another deck with the Regenstein name, this time paired with Scwartz. This deck has standard courts.

U26a

U27

U27 STRAUSS & TRIER
New York, c1860. The authors know only one deck by this maker. The court cards bear no resemblance to those of any other manufacturer of this period.

U28

U28 A. BALL & BRO.
Chicago, c1910. The New York Consolidated Card Co. probably made this Faro deck for this company. Ball was a jobber of gambling and gaming equipment which issued catalogues offering Faro layouts, dice, cheating devices, dealing boxes, card presses and playing cards. There is no evidence that they ever made any cards. Note that the King of Hearts is identical to the Hart Faro decks of the period.

Since the publication of the original Encyclopedia, we have discovered that James Y. Humphreys of Philadelphia, the renowned producer of the Seminole Wars deck, also produced a number of decks of standard cards, likely in the period from about 1810 to 1825. Until these discoveries, like Jazaniah Ford, Humphreys was thought to have produced only a commemorative deck (W2) and to have been a distributor of other makers' cards. The first mention of Humphreys was in an 1800 advertisement where he offered "an assortment of the very best English cards" and "from Boston an assortment of American playing cards". The cards were offered on a "wholesale or retail basis on most reasonable terms". By 1816 Humphreys was also listed as a card manufacturer and by 1818 was advertising "American, French and Spanish playing cards of every description and quality" from his manufactory at 86 Front St.

EARLY HUMPHREYS

U29

U29 J. Y. HUMPHREYS
Philadelphia, c1816. The one way courts in this deck are unusual, and in the English style. All the courts are hand colored, and the pips appear to be stenciled. J.Y. Humphreys' factory was located in Philadelphia, but no address appears on the Ace of Spades.

U29a

U29a J. Y. HUMPHREYS
Philadelphia, c1816. This deck, with quite a different Ace of Spades, has 13 stars like U29. The cards have bright, rather unusual coloring hand stenciled on the courts, which are very different than U29, and a very fine engraved Ace of Spades.

U29b

U30

U29b J. Y. HUMPHREYS

Philadelphia, c1816. This Ace has Philadelphia printed on it, 16 stars and the words "American Manufacture" on the banner and is more reminiscent of some of the other early makers. Interestingly, the courts are identical to those of U29a.

U29c J. Y. HUMPHREYS *(not pictured)*

Philadelphia, c1816. Another version of a Humphreys Ace with the central Spade pip on a mounted shield of stars and stripes. There is no mention of Philadelphia or American Manufacture in the banner. This Ace was a forerunner of the early Aces of L. I. Cohen.

U30 JONES & COMPANY

New York, c1860. This deck has served only to confuse the issue more. The cards show the Jones & Co. name on the Ace of Spades. The Stars and Stripes, plus the eagle at the top, strongly indicate that the cards were made in the U.S. The back could well have been designed by Owen Jones and the Ace of Spades and the King of Hearts are quite similar to those made by J.J. Levy (see NY17). There is an argument for moving this listing to Chapter 5 (NYCC) but we will leave it here until further information is forthcoming.

U29d

U31

U29d J. Y. HUMPHREYS

Philadelphia, c1816. A Spanish suited deck published by Humphreys. The wrapper for this deck has the Spanish royal coat of arms over the words "Naipes Espanoles" and the words "Manufacturado por James Y. Humphreys No. 36 Calle primera Sud, Philadelphia". The cards are interesting in the American symbolism depicted on the Aces. Note the monogram 'J.Y.H.' on the 7 of coins.

Several years ago Gene Hochman purchased a mint deck packaged in a Jones & Co. wrapper. There was part of an Andrew Dougherty private die tax stamp with no date. The deck inside, AD8, had rounded corners and was known to be c1880. Gene was sure that the backs were designed by Owen Jones. Everything indicated that the deck was made in the United States, but the wrapper indicated that it was imported from England. As Owen Jones was known to have been employed by Lawrence & Cohen in 1865 (see NY11), a question arose in his mind. Were the "Owen Jones" decks actually imported from London? Or did Owen Jones work in the United States with companies attempting to portray an imported image through the creation of fictitious wrappers?

U31 SAMUEL M. STEWART & CO.

Philadelphia, c1840. An unusual deck made by Sam.L M. Stewart. A search of the business directories failed to disclose any mention of this firm. The intricate details of the Ace of Spades indicate a date toward the middle of the 19th century, but it could be earlier. The back design was found on decks of this period made by Caleb Bartlett and others, but this does not indicate any connection with Bartlett, as paper for the back designs was sold by paper firms and not made by the manufacturers themselves.

U32

U32 ROBERT SAUZADE
New York, c1830. This maker of early playing cards had been found listed in directories but no examples were known. It has recently shown up in the Bollhagen Collection. The Ace of Spades is just a plain black pip in the center of the card making identification from that source impossible. The courts, however, contain either the name or initials of the maker in every instance. This particular deck is quite worn, and unless one is aware of the fact that the name or initial might exist on the courts, they could easily be overlooked.

U35

U35 CARMICHAEL, JEWETT & WALES
c1840. This maker was also previously unknown but two decks have surfaced relatively recently. This full size deck has an elaborate Ace of Spades and uncommon single ended courts.

U33

U33 JOHN CASENAVE
New York, c1805. Another manufacturer of early playing cards who was listed in the original Encyclopedia on the page of 'unknown makers'. Subsequent to publication, Gene Hochman found a wrapper bearing his name that was framed with an Ace of Spades and several courts. This identified an anonymous deck that he had in his collection. The maker's name had not appeared anywhere on the deck, only the words "Fine Playing Cards, New York". Another mystery solved!

U36

U36 CARMICHAEL, JEWETT & WALES
c1840. A completely different Ace for this miniature deck measuring just over 2 inches by 1 1/2 inches. The courts are very similar to those of U35.

U34

U34 EMPORIUM
J. & L. D., c1840. This deck has a finely engraved Ace of Spades and unusual one way courts. The banner says "E Pluribus Unum" instead of the more common "American Manufacture" and the initials J. & L. D. appear under the banner. Were these the initials of the maker? Was Emporium a store or some type of retail outlet? We have tried, but been unable to attribute the deck to a known maker. Another mystery!

U37

U37 THE STERLING CARD COMPANY
24 Vesey Street, New York, c1876. This deck seems to have been designed by a completely unknown maker, as the courts are not reminiscent of any other decks we have seen. It is a high quality product with a fine finish and gold edges. The deck was obviously someone's attempt to come up with indices that would be acceptable to the card playing public at a time when much experimentation was going on. The indices have some similarities with those used by the Eagle Card Co. on L5. We know of only two of these decks and neither has a Joker (which might be expected around this date) although we know of many decks from around 1875 that were issued as 52 card decks or with a blank. In addition, as both decks are mint in their original boxes, it is quite possible that they were issued without Jokers. The box states "patented on October 17, 1876 by Cornell & Shelton in Birmingham, Ct."

U38

U40 GEOGRAPHY DECK
William Montagu, 1793. This geography deck is very similar to a deck produced in England by a relative of Montagu. It is the only American deck of this type that we are aware of and certainly one of the very earliest decks made in the United States.

U40

U38 DECATUR VICTORY
Hunt, c1815. An unusual deck which has the name of a well-known English manufacturer. Was it made by the English firm for export to the United States or by an American firm of that name? The Ace of Spades depicts the Decatur Victory of the early 1800s.

Henry Hart produced cards in the latter portion of the 18th century in England and his cards, with exportation Ace, have been found in the United States. The wrapper listed as U39 and shown below indicates he also produced cards in America, or were there two Henry Hart's? Or did someone import Henry Hart decks and put them in American wrappers, perhaps with an American Ace of Spades? Another mystery to be solved!

We conclude this chapter by picturing a number of early American decks about which very little is known. None of them can be properly attributed to any specific maker. Hopefully as more information becomes available, future collectors will be able to solve their identities.

U39 HIGHLANDERS
Henry Hart, c1810. While we have not been able to locate a copy of this deck, we assume it exists because of the pictured wrapper.

U39

LONGLEY BROTHERS AND SUCCESSOR COMPANIES

The name Longley is not well known amongst playing card collectors, likely as very few decks bore their name, but the Longley brothers, and the companies they were associated with, produced cards for over 50 years using a variety of names and styles. Many examples of the cards these companies produced survive to this day.

Abner Longley, an itinerant preacher and sometime cabinetmaker in the first half of the 19th century had seven sons, five of whom entered the printing business around 1850 in Cincinnati. When they disbanded in 1860, two of them struck out together to manufacture flags and paper goods, and manufactured their first deck of playing cards in 1861. For the next 40 years or so, members of the Longley family were directly involved in the manufacture of playing cards in a variety of different locations and under a variety of trade names. From the early 1900s on, their main involvement appears to have been with American Playing Card Company of Kalamazoo, Michigan that ultimately fell into the USPC fold sometime around 1915.

The history of the Longley companies and trade names owes much to the research of the late Theodore A. Longstroth, a collector and researcher from Cincinnati, Margery Griffith, for many years the curator of the collection of the United States Playing Card Company, and Ron Decker, the current curator. Most of the information is based on fact, some on deduction and a bit on speculation (ours). In many cases, the fact that a particular company was connected to the Longley's is clear. However the timing and circum stances of the connection are murky. For example, Union Playing Card Company was certainly a Longley company, but did they start it, acquire it, merge with it? We do not yet know. No doubt, the future will shed further light on this and other aspects of this interesting group of playing card companies.

The Longleys originally manufactured cards in Cincinnati and while they then started manufacturing in Middletown, Ohio

(about 30 miles from Cincinnati) in 1867, they appear to have also continued in Cincinnati for some years after that date. The first deck produced by Servetus and Septimius Longley in 1861 had the name Longley Brothers Playing Cards on the Ace of Spades. From that simple beginning, the Longleys' association with playing cards spread widely and the following is our attempt to list the many companies and brands that stem from the Longley connections, in chronological order. We recognize that we may have included a few of the smaller companies in error, that there are overlaps and likely some omissions, and that exact dating information is sparse.

Longley Brothers Playing Cards, Cincinnati	1860s
Cincinnati Playing Card Company, Cincinnati	1860s
Fabrique Card Company, Cincinnati	1861
Eagle Playing Card Co., Cincinnati (Samuel Cupples, sole agent)	1866
Cincinnati Card Co., Longley & Brothers, Mfg., Cincinnati	1867
American Card Co., Cincinnati	1860s to 1870s
Globe Playing Card Company, Middletown (Ohio)	1860s to 1880s
Paper Fabrique Company, Middletown	1860s and 1870s
American Playing Card Co., New York	1870s
Card Fabrique Company, Middletown	1870s and 1880s
Eagle Card Co., Middletown	1860s to 1880s
United States Card Co.	1870s
Union Club Card Co., Middletown	1880s
Excelsior Card Company, Chicago	1880s
Columbia Playing Card Co., New York	1880s and 1890s
Card Fabrique Co., New York & Chicago	1880s and 1890s
Globe Playing Card Company, New York & Chicago	1880s and 1890s
The Union Playing Card Company, New York	1870s to 1890s
The Empire Card Company, New York	1880s and 1890s
American Playing Card Company, Kalamazoo (Michigan)	1880s to 1915
Paper Fabrique Company, Basic City (Virginia)	1890s and 1900s
The Eureka Playing Card Co., Chicago, St. Louis and Detroit	1890s
Knickerbocker Playing Card Co., Albany	1890s
Brooklyn Playing Card Co., Brooklyn, N. Y.	1890s
Chicago Card Co., Chicago	1890s and 1900s
Chicago-American Card Company	1890s
North American Card Co., Chicago	1890s and 1900s
Layton & Co.	1890s
The Bay State Card Company	1890s and 1900s
Union Card Co., Chicago	1890s
Wm. Hobkirk & Co., Chicago	1880s
M. F. Milward, Chicago	1890s
The New Chicago Playing Card Co., New Chicago (Indiana)	1902 to 1905
The Chicago Playing Card Co., Waukegan (Wisconsin)	1903
Detroit Card Company, Kalamazoo	1900s

The listings in this chapter include items from four different sections of the original Encyclopedia, listings U20 to U25 from Early Makers and MSW1 to MSW18, MSW53 to MSW69 and MSW123 to MSW130 from the Other Makers. To this nucleus a few other items have been added from the original work and a number of new items have been listed for the first time.

The order of listing in this chapter cannot follow strict chronological lines but does group the various decks in rough order starting with the Cincinnati production, then Middletown, then the Chicago area and, finally, American Playing Card Company of Kalamazoo. A new Hochman identification code ('L') for all these decks has been used.

L3 GREAT MOGUL
Paper Fabrique Company (formerly U21), Middletown, c1867. An identical deck to L1 and L2 with a different name.

L3

L1 GREAT MOGUL
Cincinnati Card Company, Longley & Bro., Manufacturers, No. 143 Walnut Street, Cincinnati, 1866. Through trade directories it has been established that the Longleys were only at the Walnut Street address in 1866/67.

L1

L4

L4 STEAMBOATS #2
(Formerly MSW53), Eagle Card Co., Middletown, Ohio, c1870. This Ace of Spades is clearly the same as those of L2 and L3, The American Card Co., and Paper Fabrique Company decks, except it is unnamed. This deck also had a version of the Heathen Chinee Joker, made popular by the early Eagle packs. The Eagle Card Co. name on this deck appeared on the wrapper only.

L2

L2 GREAT MOGUL
American Card Co. (formerly U20), Cincinnati, 1862. The wrapper for this deck mentions "Longley & Bro." and has a five cent tax stamp canceled 1862.

L5

L5 EAGLE CARD CO.
(Formerly U22), Cincinnati, c1867. In addition to the name of the company, this deck refers to Saml. Cupples & Co. of St. Louis as Sole Agent. The deck has no indices, one-way courts and an early Joker. It is unclear where Mr. Cupples' territory was – perhaps just St. Louis or west of the Mississippi – but it probably did not include the Eastern United States. Mr. Cupples was formerly from Cincinnati and was a friend of the Longley family. Interestingly, one of his other business ventures in St. Louis was the Cupples Envelope Co. which was the publisher of the St. Louis World's Fair (Louisiana Purchase) deck listed as SX21.

L6

L6 EAGLE CARD CO.
(Formerly U23), Middletown, c1870. This deck features a Joker representing Bret Harte's 'Heathen Chinee' (similar to L7 Joker). Certain of the pip cards were specially marked for the game of Casino (two of Spades "Little Casino" and 10 of Diamonds "Big Casino") and some for the popular game of the day called Pedro (the three's are "Don", the five's "Pedro" and the nine's "Sancho"). The deck has square corners and no indices with two-way courts.

1870S TRADE CARD

L7

L7 EAGLE CARD CO.
(Formerly U24), Middletown, 1877. In 1877 the patent mentioned as "applied for" on the Ace of Spades of L6, was granted and an additional patent was received for a new type of indices, which in all respects other than the new indices, is the same as L6.

The American Playing Card Co. of New York made several decks, c1875, before they disappeared. Research has shown that there is a marked resemblance between the King of Hearts in the early American Playing Card Co. of Kalamazoo (definitely a Longley association) decks and the next few listings for American Playing Card Co. of New York. Another indicator of the relationship with other Longley companies is the use of the American name which they had used early on in Cincinnati. In any event, because of the likely connection, the American Playing Card Co. of New York decks are listed in this chapter.

L8

L8 UNITED STATES CARD CO
New York, c1880. Based on comparisons of the courts and the Joker with the Eagle Card Co. deck L5 we can be sure that this brand was another Longley foray into the Eastern Market.

L9

L9 AMERICAN PLAYING CARD CO.
(Formerly MSW22), New York, c1875. Probably the earliest of the decks listed by this maker.

L10

L10 AMERICAN PLAYING CARD CO.
(Formerly MSW23), New York, c1876. A beautifully made deck with gold edges, which came in a sturdy telescope box. The Joker is quite similar to L9. The courts (King of Hearts shown) are an unusual style.

L11

L11 BROADWAY STEAMBOATS #288
Columbia Playing Card Co, New York, c1876. A similar Ace of Spades to L10 but with a different company name and different courts. The Joker is also similar. The quality of this deck is similar to L10a. Refer also to L81 and L82 for two other decks made by this company.

L10a

L10a NEWPORT STEAMBOATS
American Playing Card Co., New York, c1876. A much cheaper quality deck with the name on the box. The box is identical, except for the name, with that of L11.

L12

L12 AMERICAN PLAYING CARD CO.
(Formerly MSW24), New York, c 1875. "Our celebrated CONFECTIONERY PACKAGE.... containing one pack of our Round Cornered, Indexed, Playing Cards." This is the description on a box, wide enough to contain two packs, but perhaps containing one along with some candy or sweet. The courts and the printing indicate the deck was made by American PCC, New York.

L10b

L10b AMERICAN PLAYING CARD CO.
New York, c1876. Pictured is a another Ace of Spades used by this company.

L13

L13 CENTENNIAL PLAYING CARDS
Longley & Bro. Wholesale Manufacturers, Cincinnati, 1876. The special deck to commemorate the centennial of the Revolutionary War was conceived by the Longley's in the years leading up to 1876. Seventy-five metal plates, mounted on cherry wood, were created to produce 52 cards in red and blue on a white background. The suits are Crossed Swords, Bugles, Cannons and Anchors. The courts are all Revolutionary War heroes. This deck was never issued and only a few packs were produced as examples. Several cards from one of these are in the USPC collection.

L14

L17

L14 AMERICAN CLUB #165, BORDER INDEX CARDS
Paper Fabrique Company (formerly U25), Cincinnati, 1878. Another patent for a new form of indices was received on December 25, 1877. Interestingly, the holder of this patent was Cyrus W. Saladee, the holder of the patent on the scarce 1864 Samuel Hart deck known as Saladee's Patent (NY44). In this deck the Joker represents a Chinese man holding cards and the courts depict 12 members of royalty then reigning in Europe.

L17 EUCHRE
(Formerly MSW54), Card Fabrique Co., Middletown, c1880. This beautifully engraved Ace of Spades was used in this special Euchre deck as well as their regular 52 card decks of this period. The game register shown here was packed with the 32 card Euchre decks.

L15

GAME REGISTER

L15 WHIST AND POKER, #175, PATENT DOUBLE SYMBOL CORNER INDEX CARDS
Paper Fabrique Co., Cincinnati, 1878. Another new form of indices again patented by Saladee. These were heralded as being "new, neat and saleable and certain to supersede the old hackneyed styles and become the cards of the future".

L17a BOUDOIRS (not pictured)
Card Fabrique Co., Middletown, c1880. A beautiful deck with substantial use of gold on the courts, back and game register, which utilized the same Ace and Joker as L17. The picture on the back matches the register and the box has a picture of the Joker stamped in gold on a blue background.

L16

L16 ROYAL FLUSH
(Formerly MSW57), Card Fabrique Co., Middletown, c1875. This deck has an unusual looking Ace of Spades printed in deep blue. This is the Ace held in the hand of the Jokers in L17 and L18. It is also feasible to assume that a similar Ace of Spades was used for a Globe Playing Card Co. deck.

L18 UNION CLUB CARD CO. (not pictured)
(Formerly MSW55), Middletown, c1880. The same Ace of Spades as L16, with a different Company name. The tiny print on the Joker (see L17) still reads "Card Fabrique Co., Factory, Middletown".

L18a

L18a B. P. GRIMAUD
France, c1890. We have included this deck by the well known French manufacturer as the Ace of Spades is the same as the last two listings, except for the four corner indices. Does this Ace and the use of a French word like 'Fabrique' indicate a connection between Grimaud and the Longleys? Perhaps the Longleys obtained know-how from Grimaud or perhaps Grimaud had lent money to or invested in the Longleys ventures? This deck, while stamped on the Ace of Clubs with the French tax stamp, was found in the United States and may well have been produced for use in America. The courts are very much in the American style.

L20

L20 4-11-44
The Globe Playing Card Co., Middletown, c1885. This deck was named after a popular card game of the era and again used the 'hand holding cards' Joker of L19 with the Globe name.

The Card Fabrique Company with participation by Servetus, Septimius and Servetus' son Herbert operated to approximately 1894. The Globe Playing Card Co. (Servetus and Septimius' son William) operated from about 1889 to 1894. Both companies produced cards from the Middletown plant. In addition to standard brands, they also made a number of advertising decks and a Lion Brand deck advertising The Wind'sch-Muhlhauser Brewing Co. of Cincinnati is a fine example.

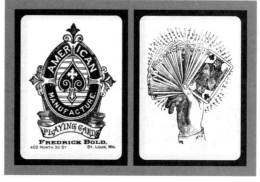

L21

L21 FREDERICK BOLD
St. Louis, Missouri, c1885. Based on the Ace of Spades the deck was made by Globe of Middletown for Frederick Bold whom we believe supplied magicians (and maybe gamblers!) with playing cards. A later listing in this chapter (L78) for F. Bold of St. Louis is for a 'wedge' deck and clearly made by American PCC of Kalamazoo.

L19

L19 STEAMBOATS #5
(Formerly MSW56), Globe Playing Card Company, Middletown, Ohio, c1880. The Joker is the same as L17 above, but the small print is changed to read The Globe Playing Card Co., Factory, Middletown.

In 1886 The Card Fabrique Co. announced to the trade that they had "established Mr. W. H. Longley as our representative in Chicago with headquarters at 142 Dearborn Street, Room 19." Other ephemera of that time refers to W. H. Longley as "North Western Manager" and indicates that manufacturing remained in Middletown. In addition, based on a Globe PCC price list dated 1891 we know that manufacturing was still being done in Middletown. We will also come across William Hey Longley again in connection with American Playing Card Co. of Kalamazoo, Michigan later in this chapter.

L22

L22 4-11-44
(Formerly MSW58), Card Fabrique Co., New York and Chicago, c1886. The brand name was on the box only. Another brand using this Ace was called 'The Elks'. The Ace and Joker are the same as L20 with the Card Fabrique name.

L22a 4-11-44
(Formerly MSW59), Globe Playing Card Co., New York and Chicago, c1890. The same deck as above, but found with no box. The Joker is the same but with the Globe name.

L23a

L23a CHICAGO-AMERICAN CARD CO.
Chicago, c1895. This deck, another mystery, was likely manufactured by a Longley company. Another deck with this Ace (Steamboats #2222) had the pictured Steamboats Joker.

To conclude with the Middletown period, a price list for Card Fabrique Co. from 1886 listed the following brands:

Steamboats #4	Steamboats #5
Lions #10	Traveler's #15
4-11-44 #18	4-11-44 #19 (Gold Edges)
Highlanders #20	Sportingmans #30
Imperials #35	Elks #40
Ideals #45	Boudoir #50
Extra Elks #60	Extra Boudoir #70
Progressive Euchre or Besique #80	Extra Progressive Euchre or Besique #90

The following decks from the Chicago area, manufactured in the late 19th or early 20th centuries, were likely made by one of the Longley companies.

L24

L24 ROUGH RIDERS #90
(Formerly W43 and MSW69), Chicago Card Co., Chicago, c1895. A Teddy Roosevelt deck with war type scenes on the box, back and Joker. The courts were the same as L25. This Ace was also used for other brands.

L23

L23 CLIMAX #55
Union Card Co., Chicago, c1895. The only deck we have seen by this specific Union company. There are some similarities to other makers in this section and we speculate it was another name used by Longley in its Chicago days.

L24a CLIPPERS
Chicago Card Co., Chicago, c1895. Another Ace used by this company.

L24a

L25

L25 GOOD LUCK #120
Chicago Card Co., Chicago, c1895. The courts were one of two unusual varieties used by this company and identical to those of the Excelsior version of Good Luck #120 listed as L29a. The pictured Joker is also unusual.

L26

L26 STEAMBOAT #222
North America Card Co., Chicago, 1897. This deck, with the identical courts to L25a, was also sold as Chicago Steamer #222.

L25a

L25a FLORAL BACK
Chicago Card Co., Chicago, c1895. The same Ace of Spades but with a different Joker and another set of unusual, striking courts. These courts are the same as the next two listings for North American Card Co. Backs included Daisy, Carnation, Forget Me Not, Lily of the Valley, Apple Blossom and Violet.

L27

L27 AMERICAN BEAUTY CARDS #444
(Formerly MSW103), North American Card Co., Chicago, 1897. This deck has the L26 Ace and the pictured Joker. It was also issued under the brand name Anchor.

L25b

L25b CRESCENT CARD CO
Chicago, c1905. Was this the same company? Evidence indicates the courts are the same as those in L25.

L27a

L27a XRAY #555
North American Card Co., Chicago, 1897. The same Ace as L26 except it has the number '555' directly under the spade.

L28

L28 LAYTON & CO.
Chicago, c1895.

There are enough similarities between the preceding three decks and some of the other Longley decks for us to include them in this chapter. L28a and L28b might have been manufactured for stationers or department stores in the Chicago area.

Finally, we have a later deck by Western PCC with almost the same Ace (L28b) that Gene Hochman had dated about 1915.

L28c WESTERN PCC *(not pictured)*
(Formerly MSW137), Chicago, c1915.

It also appears that Bay State Card Company had a connection with Card Fabrique, Globe, North American Card Co., etc. as the earliest Ace of Spades is clearly derived from L22 and the courts are the same as those of L27. Whether one of these companies was sold to someone who started Bay State, or whether Card Fabrique/Globe started it is unclear. By comparing courts we are sure that Bay State became part of Standard Playing Card Company of Chicago in the early 1890s and thereby part of the USPC fold in 1894. We have seen Bay State decks with Standard PCC, NYCC, USPC and Russell PCC tax stamps. (The connection with Russell was demonstrated by Gene Hochman in Volume VI of the original work.) We have listed the Bay State decks with Standard in Chapter 12.

L28a

L28a NOVELTY PLAYING CARDS
Wm. Hobkirk & Co., Chicago, c1880.

L29

L29 STEAMBOATS #66
(Formerly MSW80), Excelsior Card Co., Chicago, c1895. Careful examination of the Joker from this Steamboat deck will reveal that the hand is holding an Ace of Spades similar to the Card Fabrique deck, L17. Clearly this deck was made by Card Fabrique/Globe at the time they were operating from Chicago & New York.

L29a GOOD LUCK #120 *(not pictured)*
Excelsior Card Co., Chicago, c1888. The same Ace as above but the exterior of the box and the courts are identical to those of the Chicago Card Co. deck L25a. Unfortunately our copy does not have a Joker which, if like L17, would have added further proof of the connection.

L28b

L28b NOVELTY PLAYING CARDS
M. F. Milward, Chicago, c1890. Two different Jokers that we have seen with this Ace are pictured.

L30

L30 EXCELSIOR PLAYING CARD CO.
(Formerly MSW81), New York, c1890. The Joker of this deck was the first to feature the card-playing dogs that became so popular several years later.

L31

L31 EXCELSIOR PLAYING CARD CO.
(Formerly MSW82), New York, c1890. Another deck made by the same company in the same time period with the same Joker.

The Joker with the last two decks has an emblem which seems to say 'CB Co.', likely a reference to Caterson & Brotz Co. Was this Philadelphia manufacturer related to the Longley companies? While comparisons of their court cards and other features with those of contemporary and early 20th century makers produce few similarities, there are a number of reasons to suspect so, and we have therefore included the Caterson & Brotz listings in this chapter. Hopefully one day more definitive proof of the connection will emerge. In the meantime we note the CB Co. on the Joker; a remarkable similarity between a Caterson & Brotz deck (L33) and an early deck by American Playing Card Co. of New York (L10); and the connection between American of New York and American of Kalamazoo described earlier in this chapter.

Caterson & Brotz produced a number of decks in Philadelphia, from about 1880 until the turn of the century when they disappeared from view. Known brands, from an 1888 price list, include:

Canoe #26	Socials #36
Dom Pedro #56	League #58
Bon Ton #66	Peanuckle #60
Euchre #73	Solo #75
Grand Duke #76	Embassy #78
Peanuckle #80	Imperial #96
Emperor #98	Triumph #106

L32

L32 LEAGUE
Caterson & Brotz Co. (formerly MSW63), Philadelphia, c1880. This deck appears to be their earliest. The Joker, marked 'Cat' likely indicated a nickname for Caterson. This deck was also found with the other Jokers pictured – on one a bull is seen derailing a train and on the second (named 'League') a baseball player seems to be catching a ball with his teeth. This Ace of Spades was also used for several advertising decks, always with a special advertising Joker.

L32a

L32a CHIP CARDS
(formerly MSW67), Caterson & Brotz Co., Philadelphia, 1888. Another version of this deck was made for commuters and other railroad travelers. The patent date on the box is 1888 and the deck came packaged in a nice box, with a steam engine design. The backs show the inside of a parlor car with 'games' in progress. The box also contained 96 paper chips, in three different colors, which were punched out of eight sheets of cardboard. It had yet another Joker.

L33

L33 CATERSON & BROTZ CO.
(Formerly MSW64), Philadelphia, c1885. A new Ace of Spades which, as mentioned above, is virtually identical to one by American Playing Card Company of New York. This brand featured the novelty back designs becoming popular at that time with the larger manufacturers.

L35a

L35a IMPERIAL #96
Caterson & Brotz Co., New York & Philadelphia, c 1885. A completely different version of this brand with a Best Bower 'Monk Joker'. The back design has a similar design to the Ace of Spades. This deck was also issued under the brand name Emperor.

L34

L34 ROUGH BACKS
(Formerly MSW65), Caterson & Brotz Co., N.Y. & Philadelphia, patented March 9, 1886. Quote from the box: "Rough Backs prevent misdeals. These cards are roughened on the backs to cause an air space as well as to make resistance for the thumb in dealing, and highly finished on the face to avoid too much resistance. They slide off one by one, and do not become sticky in any weather, and improve by use." Note the similarity of the Joker to that of L10.

L36

L36 NOVELTY
Caterson & Brotz Co., c1885. This deck came with the 'Baseball' Joker pictured under L32.

We should also mention that Caterson & Brotz made a number of early advertising decks including the well known 5A Horse Blankets and Edmund's Cigars decks.

L35

L35 IMPERIAL #96
(Formerly MSW66), Caterson & Brotz Co., N.Y. & Philadelphia, c1885. The box reads: "Caterson & Brotz, Philadelphia, New York Office at 129 Crosby St., Leo Schlesinger, Manager." The Joker was the same used in L34. We have also seen this deck with the Union Joker used for L40.

L37

L37 REYNOLDS CARD MANUFACTURING CO.
(Formerly MSW68), New York, c1882. Another 'unique' deck and maker. The similarities to Caterson & Brotz are so strong that we can safely assume that it was made by them.

L37a

L37a ROUGH BACK
The Reynolds Card Mfg. Compy, New York, 1885. Another Reynolds deck with an unmistakable connection to Caterson & Brotz through the name "Rough Back" and the Joker from L34.

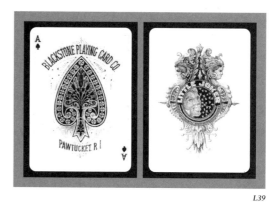

L39

L39 BLACKSTONE PLAYING CARD CO.
(Formerly MSW48), Pawtucket, R.I., c1885. Comparison of the Blackstone decks indicates a relationship with Card Fabrique.

L38

L38 BROOKLYN PLAYING CARD CO.
(Formerly MSW51), Brooklyn, New York, c1896. This deck was found in the U.S. Playing Card Co. collection in Cincinnati. The Ace of Spades is similar to one by Knickerbocker Playing Card Co., of Albany, N.Y. (see L38a below). Research at the Museum has led to the conclusion that Brooklyn was a name used by Card Fabrique Co.

L39a

L39a BLACKSTONE PLAYING CARD CO.
(Formerly MSW49), Pawtucket, R.I., c1885. A very different Ace by this company. No Joker was found with the deck.

The Union Playing Card Company of 79 & 81 Duane Street, in New York, was another of those firms with many names. Cards that can be definitely identified as theirs have been found under the Empire Card Co., the Eureka Playing Card Company and the Universal Playing Card Co. They produced a varied line and made cards from about 1875 until the mid-1890s.

L38a

L38a KNICKERBOCKER CARDS
(Formerly MSW93), Knickerbocker Playing Card Co., Albany, N.Y., 1890. The same Ace of Spades as the Brooklyn Playing Card Co. If Brooklyn was part of Card Fabrique Co., it is likely that Knickerbocker was connected as well.

L40

L40 THE TRAVELER'S COMPANION
(Formerly MSW123), The Union Playing Card Co., New York, 1886. In addition to the regular deck of 52 cards and Joker, this package contained 2 rolls of small poker chips for play on trains, etc.

L41

L42

L41 UNION PLAYING CARD CO.
(Formerly MSW124), New York, c1875. A regular 52 card deck with an interesting early Joker. Note the initials U.P.C. Co. in the center of the Ace.

L42 SQUARED PHARO
The Union Playing Card Company, New York, c1880. A deck made for the Pharo players who preferred the traditional square cornered cards with single ended courts.

EARLY UNION PCC TRADE CARD

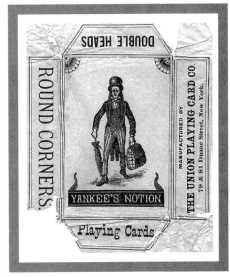

UNION WRAPPER

L43

L43 UNION PLAYING CARD CO.
(Formerly MSW125) New York, c1880. A different Ace of Spades, but the U.P.C. Co. initials are still prominent. Pictured are three different Jokers we have seen used with this Ace of Spades. The deck with the Heathen Chinee Joker came in the wrapper with the brand name Yankee's Notion.

L43a

L43a CONCORDANT CARDS

Concordant PCC, New York, c1885. The similarity of this Joker leads us to suspect that this otherwise unknown company was somehow related. The name only appeared on the box.

L43b

L43b GRAPHIC CARDS

Graphic Playing Card Co., New York, c1885. This deck has the identical Ace of Spades and Joker as L43a but the box (pictured) states it was made by the Graphic Playing Card Co., New York. The box and texture of the cards indicate Union PCC made it, either for Graphic PCC or for Union to use in a different market.

L44

L44 BERNARD DREYFUSS, SOLE PROPR.

The Union Playing Card Company, New York, c1880. The same Ace as L43 with the words Bernard Dreyfuss, Sole Propr. added. Who was Mr. Dreyfuss? We do not know but, based on the rarity of decks with his name, we can assume that Union did not make this deck for a very long period. This deck came with the Best Bower Joker shown with L43.

L45

L45 THE EMPIRE CARD CO.

(Formerly MSW126), New York, c1880. A very similar Ace of Spades to that of L44 fronting an identical deck. Although the overall design of the Ace is completely changed, one has little doubt that it was made by the same firm, especially as it has the same Best Bower Joker as L43.

L46

L46 SPORTING CARDS

(Formerly MSW127), The Union Playing Card Co., New York, c1885. No doubt made to compete with the USPC's Sportsman brand, this deck has a sporting Ace and back plus a humorous Joker.

L47

L47 SPORTING CARDS

(Formerly MSW128), Empire Card Co., New York, c1885. Again a deck so similar, it is easily recognized as being by the same manufacturer.

L48

L48 EUREKA PLAYING CARDS

(Formerly MSW129), The Union Playing Card Company, New York, c1895. A new Ace of Spades with the Heathen Chinee Joker used in some versions of L43.

L48a

L48a EUREKA PLAYING CARDS
c1895. Another version with the maker's name left off, no doubt due to the very poor quality of this deck. The known copy is missing the Joker but it was likely a version of the Heathen Chinee.

L49

L49 EUREKA PLAYING CARDS
(Formerly MSW130), The Eureka Playing Card Co., Chicago, St. Louis, Detroit, c1895 with a "Here I am again" Joker. Likely the Eureka name worked well as this version indicated a new company was set up under this name.

The following information is from an insert card found with a deck of Union PCC 'Pointers', c1885.

"These CARDS ARE ALL ROUND CORNERED and the Face Cards in Each Grade are in FIVE (5) COLORS"

Yankee's Notion, Plaid Backs

Lion's, Superior Quality, Calico Backs

Rajahs, TINTED BACKS, High Finish

6. **Acme**, Enameled, Handsome Set-backs

Pioneer, Double Enameled, TINTED BACKS

5.5. **Great Mogul**, Linen, Enameled, Fancy Set-backs

4.4. **Sporting**, Enameled, appropriate backs

Extra Travellers, Enameled, Superfine Linen, Elegant Backs

Palace Car, Gilt Edges, Superfine Linen, Elegant Backs

6.6. **"Winning" Poker**, enameled face, Extra Finish

7.7. **Club**, Enameled Face, Extra Finish

9.9. **Pointers**, Pure Linen, Elegant Assorted Backs

Green Room, Gilt Edges, Elegant Tinted Backs

Progressive Euchre and Solo

Pinochle, 64 Cards

Also **Acme**, **Pioneer** & **Sporting** cards with gilt edges.

In 1891, Septimius and Servetus established the first playing card company in the South, in Basic City, Virginia. Share certificates from The Paper Fabrique Company were issued to them in March of that year. The only Ace of Spades we have seen from this location is listed below but we understand that there were several decks from this factory, although it is possible that they used the Middletown Aces of Spades for many of them. The courts were the ones originally designed in Middletown and also used by the American Playing Card Company of Kalamazoo.

L50

L50 PAPER FABRIQUE CO.
Basic City, c1891. This is an advertising deck for Hancock's Old Port (chewing tobacco) with a standard Ace and Joker.

In 1903, the Longleys were involved with a new company in New Chicago, Indiana, called New Chicago Playing Card & Flag Mfg. Co. (the Longleys had long manufactured flags along with paper goods). W.S. Longley was a major shareholder and Servetus Longley was listed as a Vice President. The operations were soon transferred to Waukegan, Illinois, just north of Chicago. The deck listed below is the only one we have seen from this company although we suspect they continued to make other brands there as well.

L51

L51 STEAMBOATS
New Chicago Playing Card Co., Waukegan, Ill., c1903. Note that this deck has the same Joker as L50.

Both L50 and L51 were rather cheap, poor quality decks. It appears that these companies were set up to service a market not covered by The American Playing Card Co., discussed on the next page.

PRICE LIST HEADER

The American Playing Card Co. seems to have started making playing cards in the 1880s in Kalamazoo, Michigan. The name, American Playing Card Co., was likely revived from the early Cincinnati days. While we believe the Longleys were involved from the beginning, we can be sure of William Hey Longley's participation from 1891 to 1913. W. H. Longley is shown as Secretary or General Manager on letterheads over a number of years, and the USPC Collection has a letter from Septimius on APCC stationary to Servetus in Basic City, Virginia, dated August 1, 1902. Finally, a letter dated April 15, 1893, on American Playing Card Co. letterhead (W.H. Longley Manager) starts off as follows:

"DEAR SIRS

Having recently purchased of the Globe Playing Card Company the right to use all of their numbers, brands, and trade marks, we start with the issue of the well known brand of '4-11-44'. We enclose herein samples of same, and desire that you will give them careful examination....."

This letter seems to confirm that Globe had closed its Middletown operations by that date and that it was transferring the rights to all its brands to APCC, no doubt because of the family connections.

The American Playing Card Co. through much of its existence also had a sales office in New York, initially at 338 Broadway and subsequently at 82 Duane St., remarkably close to the address of The Union Playing Card Co. at 79-81 Duane St. By 1908 it also had a sales office in San Francisco at 2751 Clay St. and one in Chicago in the Marquette Bldg. Records show that APCC flourished in Kalamazoo until about 1915 when many people in that area believe they combined with Kalamazoo Playing Card Co. However, a study of the history of KPCC makes this unlikely as KPCC moved to New Jersey in 1912 where they merged with Russell Playing Card Co. and evidence has APCC making cards in Kalamazoo as late as 1915. Furthermore it was recently discovered that APCC was eventually absorbed into USPC as that company is still obligated to print and sell a small number of decks annually under certain

names, in order to maintain copyright privileges. Some of the Longley trade names, especially those from American Card Co., are among the ones being maintained.

The American Playing Card Co. turned out to be a significant maker in its own right, with a large number of different brands manufactured over its existence, including several new brands discovered since the first edition. The entries L52a to L69 replace the entries in the original Encyclopedia numbered MSW1 to MSW18. Where there is a gold edged version of a particular deck, a second brand number is given in parentheses.

L52

L52 AMERICAN PLAYING CARD CO.
Kalamazoo, Michigan, c1886. This new discovery is logically the first deck made by American Playing Card Co. (APCC) and the theme of children holding cards is reminiscent of the Card Fabrique and Globe Aces of Spades, another piece of evidence in support of a connection from the start.

L52a STEAMBOATS
(Formerly MSW1), APCC, Kalamazoo, c1887. Originally believed to be the first deck by APCC this Ace of Spades was found with a Steamboat #99 deck. So far no other decks with this Ace have been noted, but it is quite possible it was used for other brands. While no Joker was found with the deck, one was likely issued.

L52a

Other than these two Aces of Spades, only one other type seems to have been used by APCC. From about 1890 to 1915, only this third Ace was used, despite the fact that American issued many brands. Of course, minor changes were made throughout the many re-engravings made necessary through constant use over a 25 year period.

The name of the brand was usually only found on the Joker and the wrapper or box. Decks found to be missing these two important items can be identified as an APCC product, but the brand name and year of issue will likely remain a mystery!

L54

L54 FARO
(Formerly MSW3), APCC, c1890. The only square cornered deck produced by APCC. The deck has the quality and backs of the Steamboats and came in a paper wrapper. The Joker was unnamed. In an 1891 price list this deck was described "No.4 Steamboats, square corners, Star, Calico and Plaid backs, different colors. Every Pack in a Neat Tuck Box". The price was $5 per gross.

APCC FACTORY

L55

L55 EAGLES
(Formerly MSW4), APCC, c1890. This deck also had a Joker, which bore no brand name. The box said "Super Enameled", so it was probably one of their better brands. Note that this Joker bears a strong resemblance to the eagle used on the interior of the Spade Ace. Perhaps it was their first brand to use the new Ace.

L53

L53 STEAMBOATS #99 (#990)
(Formerly MSW2), APCC, c1890. The common Ace with the Steamboat Joker. A completely different box was used than for L52a, and the #99 was added. The cards appear to be of identical quality. This brand was likely the lowest priced card in the line, similar to the practice of their competitors.

L56

L56 PREMIER #50 (#550)
(Formerly MSW5), APCC, c1890. This deck was their most expensive in a 1903 wholesale price list. It was made with multi-colored, pictorial backs and from super-enameled stock.

L57

L57 GOLF #98 (#980)
(Formerly MSW6), APCC, c1895. Notice the word "Joker" appears. Even if the box and Joker are missing, the backs, of which there were many, will identify this brand. Two Jokers are shown - perhaps there were more!

L57a GOLFETTE #53 (#530) *(not pictured)*
c1900. A narrow version of L57 made for whist and bridge. No doubt the same backs were used as for Golf #98.

L58 RIVALS #15 (#150)
(Formerly MSW7), APCC, c1895. A relatively cheap brand sometimes found with a sporting type back.

L58

L59

L59 KAZOO #4
(Formerly MSW8), APCC, c1895. This was a cheap card, probably close to the bottom of the line. It seems to have been phased out quickly, because it no longer appeared on company price lists by 1901.

L60

L60 BENGALS #11 (#110)
(Formerly MSW9), APCC, c1895. Another cheaper brand, costing little more than Steamboats (L53).

L61

L61 4-11-44 #18 (#19)
(Formerly MSW10), APCC, c1895. This deck was produced on Superior Enameled Cards and was probably near the top of the line. The significance of 4-11-44 is that it was the name of a popular card game of the era. As we have seen both Globe and Card Fabrique Co. produced a deck with the same name. Here again, if a deck is found without box and Joker, identification is simple due to the back (pictured). The hand holding the cards is quite similar to several motifs, which were also used by Globe and Card Fabrique Company, but American used the left hand and they used the right hand (see L17 and L18).

L62 ROVERS #20 (#220)
(Formerly MSW11), APCC, c1898. This brand was priced in the middle of APCC's range of brands. It was made from enameled stock.

L62

L63

L63 DERBY #30 (#330)
(Formerly MSW12), APCC, c1895.
This deck was the most expensive in
Class 2 per their price list, and one of
their better decks overall. It came with
a special wrapper, within a handsome
box, and its super enameled surface
and beautifully designed back make it
a most desirable deck.

L64 SOLO #21 *(not pictured)*
(Formerly MSW13), APCC, c1895. A sealed deck of Solo cards, in its
original wrapper, with an unbroken tax stamp dated 1895, was found
with 36 cards (6 to Ace). The game of Solo, popular at the turn of the
century, is played with 36 cards and does not use a Joker.

L65 DOUBLE PINOCHLE #640 *(not pictured)*
(Formerly MSW14), APCC, c1900. A 64 card pack containing no
Joker, and packed in a slip case. Many different backs, including those
used for regular brands, are found on the Pinochle decks.

L66

L66 SPANISH PLAYING CARDS
(Formerly MSW15), APCC, c1895. This
deck was produced primarily for export
and competed with Spanish suited decks
made by the other large U.S.
manufacturers.

L67 LONE HAND *(not pictured)*
(Formerly MSW16), APCC, c1900. APCC produced a narrow standard
deck with a narrow version of their standard Ace of Spades.

L68

L68 SENATORS #40 (#440)
c1890. This brand was listed in the
1890 price list but (presumably)
replaced by Elks #40 (L69). However it
reappeared later in the decade and Elks
#40 disappeared again from the price
lists.

L69

L69 ELKS #40 (#60)
APCC, c1891. Their most expensive
deck, made from pure linen stock and
with a "Super Enameled Finish". It
came in "a plated wrapper in a silk
telescope box".

L70

L70 SPORTINGMANS #45 (#450)
APCC, c1890. This deck was no doubt
made to compete with Sportsman's
#202 (US2). It featured an elaborate
Joker and sporting motif backs. It was
among the more expensive brands in
the 1890 price list. Early editions used
the brand #30 (#303).

L71

L71 COLUMBIAN #92
APCC, c1890. A high quality brand,
no doubt made to compete with the
better USPC and National decks of
the time. This deck has the same
Joker as L54.

L72

L72 BRAILLE
c1900. This deck was found with no wrapper or box, so no brand number is known. It was marked to be used by the blind and was made from excellent quality card stock.

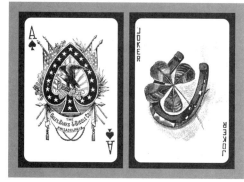

L75

L75 BAILEY, BANKS & BIDDLE
(Formerly MSN8), c1910. Another private brand, made in Whist size, similar to L74.

L73

L73 FOSTER ENGINEERING
(Formerly MSW18), APCC, c1900. APCC made a variety of advertising decks over the years. The example pictured here is from a deck made for Foster Engineering Co. of Newark. In most instances, the Ace of Spades advertises the company's product but also identifies American as the maker. The Joker is normally non-standard as well.

L75a

L75a A. I. JONES & CO. LTD.
c1910. This was another private brand designed with a golf card back for this tobacconist and stationer.

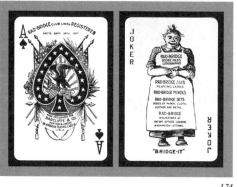

L74

L76 BROWNIE #35
c1890. This deck, with a Joker probably designed by the prolific Palmer Cox, was made in a smaller size and was "put up in handsome tuck boxes, wrapped and sealed".

L76

L74 RAD-BRIDGE
(Formerly MSW17), APCC, c1907. APCC made private brands for several companies. This one was manufactured for Radcliffe Co., of New York and London (refer AD45). The Spade Aces show the following patent dates: Club Linen, 1907; Silk Velour, 1909; and Basket Weave, 1910.

L77

L77 AMERICAN WHIST LEAGUE
c1892. A special order deck of 52 cards (no Joker) for the American Whist League. It came with a League Ace of Spades and a special back, in a Kalamazoo Folding Duplicate Whist Tray. The tray advertises many of the American brands of the day.

L77a

L77a DUPLICATE WHIST #16
c1892. Another 52 card deck made expressly for playing Whist. The deck came with the standard Ace and in the special box pictured.

L78

L78 PETER PAN
c1900. APCC also made a patience deck to compete with USPC and other manufacturers. It had the standard Ace of Spades with the pictured Peter Pan Joker.

APCC made several other brands described on the early price lists previously mentioned. No doubt they came with the same Ace of Spades and a special Joker. The ones listed are: Lions #10, Travellers #15, Columbias #105, Imperials #25 (#35), Boudoirs #50 (#70), Euchre and Bezique #90, Wolverine (Seconds), Golf Euchre #79 and Diadems #27 (#270).

L79

L79 F. BOLD
St. Louis, Missouri, c1895. A deck made for this magic supply house in St. Louis (see also L21). Comparison of the courts confirms American of Kalamazoo made it.

L80 DETROIT CARD CO. (not pictured)
APCC, c1903. This deck with the standard APCC Ace comes in a box for Detroit Card Co. #40.

The Columbia Playing Card Co. of New York, seemingly a maker of relatively low quality playing cards, was clearly a branch of the Longley companies. Its first deck (L11) is nearly identical to those of APCC of New York and the next listing has identical courts to the earliest APCC Kalamazoo decks and even the same number of stars in the flags in the similar Ace of Spades. Unfortunately this deck was found without a Joker. However, the unnamed Ace of Spades in L82 is identical to this Ace (except for the name), the courts are identical and it has the same Joker as L54.

L81 RIVALS #1110
Columbia PCC, c1890. This deck was a rather low quality brand with the same name, but a different number, as the APCC Rivals #15 (L58).

L81

L82

L82 COLUMBIA PPC
c1890. This was a higher quality deck featuring gold edges and a very interesting baseball back.

While we have tried to portray the involvement of the Longleys in the manufacture and sale of playing cards in as orderly a fashion as possible, gaps in information exist. No doubt, new Longley companies will emerge and further brands will be discovered shedding additional light on this fascinating group of playing card makers.

NEW YORK CONSOLIDATED CARD COMPANY

In 1832, Lewis I. Cohen, a stationer in New York City, made his first deck of cards. He little realized at the time that he was founding one of America's foremost companies. Indeed, his successors would form the New York Consolidated Card Company, which still manufactures playing cards today under the USPC banner.

Lewis Cohen, born in 1800, had already demonstrated that he was an enterprising young businessman. For example, he was the first maker of lead pencils in America and was also reputed to have been the first stationer in America to sell steel pens in place of the traditional quills.

The L.I. Cohen Stationery Store was located at 71 William St. and, due to the success of the playing cards, moved to larger quarters at 122 William St. in 1833. In 1835, Mr. Cohen invented a marvelous new machine, which revolutionized the entire playing card industry. His machine allowed four colors to be printed on a sheet of cards in one impression. With his secret new process, he soon achieved dominance in the market and outdistanced all of his competitors. In 1838, he moved to 118 William St. and in 1844 purchased a large building at 184-190 William St., to house his newly expanded plant. At the same time he relocated his store to a purchased building at 134 William St.

In 1854, when he was financially secure and only 54 years old, Lewis Cohen turned over control of his firm to his son Solomon L. Cohen and a nephew John M. Lawrence. The company continued under the L.I. Cohen name until his death in 1860, when it was changed to Lawrence, Cohen & Co. and, in 1864, to Lawrence & Cohen.

A second nephew, Samuel Hart, who learned the card making and stationery businesses as an employee of L.I. Cohen, left his uncle's employ and started his own business in 1844. Most of the cards bearing his name had two addresses, the Philadelphia address where his factory was located and a New York address which housed a business and sales office.

A third nephew, John J. Levy, also left L.I. Cohen in 1844 and opened a modern plant at 177-179 Grand St. in New York. He started under his own name, but soon took in a partner named Huestis, and formed a company called the Union Card Manufactory. They also made playing cards under the name Huestis & Levy. The partnership did not last too long and eventually John J. Levy was making cards again in his own name.

All three firms had access to the machine invented by L.I. Cohen, because in 1844, he decided to release the rights to his son and three nephews. This encouragement was instrumental in both Hart and Levy starting on their own before the patent rights to the invention came into the public domain.

Thus, three different firms flourished in the mid-19th century, all offshoots of L.I. Cohen, and all with the competitive advantage of the four-color playing card press. But this was not to last very long as the patent rights would expire in the early 1850s and the special machine would come into the public domain. All would soon feel the pressure of increased competition, both from each other and outside manufacturers.

Especially competitive was Andrew Dougherty, a manufacturer who was larger than any of the three individual companies. Eventually, this drove the cousins back together and in 1871 they decided to merge and reunite as the Consolidated Card Company. They gradually closed their respective plants and moved to 184 William St., the greatly expanded home of the founder. At the beginning, they continued to market the brands they were individually noted for. A January, 1873 price list for New York Consolidated Card Company at 184 William St. had three equal columns explaining that "the several interests of the firms having been amalgamated under the above title, the manufacture of their well-known Brands will be continued under the same supervision, as heretofore, and will be supplied to the Trade, at the above address, until further notice." The columns were: 1) Lawrence & Cohen (J.M. Lawrence and S.L. Cohen); 2) John J. Levy (no mention of Huestis); 3) Samuel Hart & Co. (Samuel Hart and Isaac Levy). The price list, reproduced on the next page, did not contain any NYCC brands of cards. These presumably came into their own at a slightly later date.

WHOLESALE PRICE LIST/COST PER GROSS					
LAWRENCE & COHEN	$	SAMUEL HART & CO	$	JOHN J. LEVY	$
#319 Steamboats	24	#19 Steamboats 1st quality	24	#219 Steamboats	24
321 Highlander No. 2	28	21 Highlanders No. 2	28	223 Moguls 1/2 Linen	40
318 Players	31	22 Highlanders No. 1	33	229 Moguls Dbl. Enamel	54
322 Highlander No. 1	33	23 Moguls (Fancy Backs)	40	230 Euchre deck - Heathen Chinee Joker	54
323 Fine Fancy Moguls	40	28 Barcelona	48	232 Flag Backs	54
331 Solo	48	29 Great Moguls	54	236 Independents	90
329 Moguls	54	30 Imperial Bower Euchre	54	237 Heroes	90
330 Euchre	54	32 Imperial Bower (Round Corners)	60	238 Curved Corners, fancy backs (3 colr)	96
343 Harry VIII	72	41 Imperial Eagle	72	231 Solo	48
341 Eagles	84	34 Round Cornered Whist	72		
336 Gold Backs	84	36 Gold Mogul	96		
337 Illuminated Faces	130	37 Illuminated Mogul	120		
338 Illuminated with Gilt Edge	160	38 Washington (Illuminated)	168		
340 Harry VIII Round Corners	72	Decaturs & Patent Whist	72		
342 Deutche Reichs Karten	54	Linen Eagles & Squared Faro	168		

Other early brands made by Hart/NYCC included #24 Lucky Poker, #38 Vignette and #52 Patent Linen Poker.

Certainly by 1875, newly designed brands with NYCC Aces of Spades were being brought onto the market, although it would still be many years before the Samuel Hart name would be completely phased out. The Lawrence, Cohen and Levy names must not have had as much appeal and disappeared shortly thereafter. Also, about that time NYCC moved to much larger quarters at 222, 224, 226 & 228 West 14th St.

In the Playing Card Museum of the United States Playing Card Co. in Cincinnati, is a very large drawing of the L. I. Cohen invention that revolutionized the playing card industry. It is drawn to scale and thus is too large to be given the permanent hanging place it deserves. The L.I. Cohen decks that are described and pictured below, especially if one compares the actual decks produced before and after the invention was used, demonstrate the significance of this important change in the history of this industry.

NY1a

NY1a L.I. COHEN
New York, c1832. Undoubtedly, the first L.I. Cohen issue, made without his name on the Ace of Spades. On this Ace, 'New York' is plainly written on the ribbon. The courts were identical to NY1 and this comparison, as well as comparison with later decks, positively identifies it as a product of L.I. Cohen.

NY1

NY1 L.I. COHEN
c1833. A very early deck by this manufacturer and the first Ace to bear the L.I. Cohen name. It was stencil colored and clearly made before the invention of his new machine.

NY2

NY2 L.I. COHEN
c1835. Another early deck with newly designed courts and featured 'Ivory Surface'. This was another Cohen innovation, and a forerunner of the NYCC 'Enameled Finish' which was popular at a much later date. Note the small 'LI Cohen' above the Eagle on the Ace of Spades.

NY3

NY6

NY3 HIGHLANDERS
L.I. Cohen, 190 William St., c1860. This deck still had its original wrapper and a tax stamp dated 1862. Although the Lawrence & Cohen name started about 1860, they were known to have shipped cards with the L.I. Cohen Ace of Spades. Alternatively, it could have been produced quite a bit earlier and only released from the factory in the early 1860s. The 'Ivory Surface' was still being promoted.

NY6 L.I. COHEN
c1845. A second, completely redesigned 'illuminated' deck was produced on the four color press. Although the gold leaf appears to be added by machine, it was probably done after the original printing. It is an extremely handsome deck. This Ace had newly designed courts and is also known in decks with double ended courts. It was also issued with the single ended NY5 courts (shown).

NY4

NY7

NY4 L.I. COHEN
c1840. This deck was one of the first made on the new machine. The courts were almost identical in design to NY2, but differences can be seen when compared closely. Small details in the decoration stand out which are not hidden by the stencil colorings. This deck also came with double ended courts.

NY4a L.I. COHEN *(not pictured)*
c1850. The same Ace of Spades with double ended courts.

NY7 L.I. COHEN
c1850. Although few examples remain, many of our early makers made their own version of European cards. If you note in the Huestis & Levy advertisement in this chapter, they offer 'French' and 'Spanish' cards, however none of these have been found. The deck shown here is the only example noted of an L.I. Cohen Spanish deck.

NY5

NY5 L.I. COHEN
c1840. This 'Illuminated' deck was probably the first of its type to be made in the United States. The term 'Illuminated' refers to the fact that the pips were outlined in gold (through the use of gold leaf) and gold was used extensively on the court cards. As the cards are stencil colored and the gold added by hand, the deck may have predated his machine. This type of deck was very expensive to make but, judging from the many beautiful examples by various manufacturers, they were extremely popular.

When comparing court cards from early decks minor differences will often be found. These should be ignored as plates wore out rapidly and new ones had to be engraved. Every effort was made to duplicate precisely, but as they were made at different times, and sometimes by different engravers, minor variations occurred.

We are safe to assume that all decks having a Lawrence & Cohen Ace of Spades with the William St. address, were made between 1860 and 1873. As previously noted, although organized in 1854, Lawrence & Cohen used the L.I. Cohen Aces until after 1860. In the relatively short period that Lawrence & Cohen decks were produced, their courts, both single and double ended, were used interchangeably with the Aces. For example, the same one way courts were used with NY8, perhaps their earliest deck, and with NY14 which bears the new address, 222-228 West 14th St. Demand, incidentally, for one way courts (used mainly for Faro in later years) continued into the early 20th century.

NY8

NY8 LAWRENCE, COHEN & CO.

c1860. This was probably their earliest deck with the Lawrence & Cohen name.

NY9

NY9 LAWRENCE & COHEN

c1863. A similar, but different Ace of Spades. The deck normally came with double ended courts although it is known with single ended courts like NY8.

NY9a

WRAPPER

NY9a BIJOU

Lawrence & Cohen, c1860. A patience size deck with no maker's name but clearly made by Lawrence & Cohen. It had one way courts like NY8 and came in a double box with two decks in wrappers.

NY10

NY10 ILLUMINATED

Lawrence & Cohen, c1863. A beautiful, richly 'illuminated' deck. The Ace of Spades is in royal blue and red with heavy gold accents, apparently added by hand or stencil. The crowns and objects held by the court figures are also in gold. The beautiful original wrapper stated "Superfine Florigated Playing Cards".

NY11

NY11 OWEN JONES

Lawrence & Cohen, c1865. The new firm, preparing to launch their own identity, hired Owen Jones, the famous English playing card designer who had produced many intricate and beautiful back designs in England for De La Rue. He designed new backs for Lawrence & Cohen as well as a new Ace of Spades, similar to the one used by De La Rue. Owen Jones' name appeared on the Ace of Spades (note the crown at the top of the design) and the wrappers.

NY12

NY12 LAWRENCE & COHEN

c1866. An eagle soon replaced the crown on the Aces of Spades and the Owen Jones name was dropped although the Jones backs continued to be used.

NY13

NY13 LAWRENCE & COHEN
c1871. The same type of oval design was used, but more flourishes were added in an obvious attempt to replace the Owen Jones look with the image of Lawrence & Cohen. Our copy comes with the pictured Joker which has the name Consolidated Card Co. and which also came with NY39.

NY14   NY14a

NY14 LAWRENCE & COHEN
c1872. The new West 14th Street address was used indicating it was manufactured after the formation of NYCC. This deck has the same one way courts as NY8.

NY14a LAWRENCE AND COHEN
c1873. The Ace is the same as NY14 but without the address and it has single ended courts which are like NY8, but slightly more modern.

NY14b   NY14c

NY14b LAWRENCE AND COHEN
c1873. A very different Ace of Spades with tiny indices clearly made around the time of transition to NYCC brands.

NY14c LAWRENCE & COHEN
c1865. The Ace of Spades is unidentified but very similar to NY9. The courts are the same as used for NY11 and the back design is the same as other Lawrence & Cohen decks including NY14a.

The Kings of Hearts pictured earlier were used interchangeably with Lawrence & Cohen and J.J. Levy decks. One is even found occasionally with a Samuel Hart deck. Without the Ace of Spades, there is no way to determine the true maker of one of these decks.

The cards of John J. Levy show more originality than the ones made by Lawrence & Cohen. In his first deck, he introduced a completely new set of courts. There are specimens of decks with pictorial backs, and his later decks have a fifth color added to the printing process.

NY15

NY15 JOHN J. LEVY
New York, c1845. This appears to be Levy's earliest. It was clearly done using his uncle's invention as the Ace of Spades and the court cards were done in four colors. The courts were new designs with a striking Ace of Spades. It is printed in blue, red, yellow and black. The address (in red) is hard to distinguish on the cards themselves, and nearly impossible to reproduce in a black and white photograph. The address, 177 & 179 Grand St., N.Y. is printed on the shell at the bottom of the Ace.

NY16

NY16 JNO. J. LEVY
177 & 179 Grand St., New York, c1860. This was the most common Levy Ace. You will see it twice more, with changes of address. In this example 'John' was abbreviated to 'JNO.' All of the Aces of this type were made with both one way and two way courts. This was not long after Jokers were first introduced, and this pack features a clever 'Heathen Chinee' Joker, perhaps a racial implication of deceit in gambling with cards.

NY16a

NY16a EAGLE MANFG. CO.

Philadelphia, c1865. The deck seems to have been another name used by Levy and perhaps actually manufactured by Hart for Levy if it was really made in Philadelphia.

NY17a

WRAPPER

NY17a STEAMBOATS

c1860. A single ended, generic deck presumably made by Levy based on the nearly identical Ace of Spades but without the name.

NY16b

NY16b AMERICAN MANUFACTURE

New York, c1868. We have seen two versions of this deck, again clearly made by Levy. They both have indices and one has the pictured Joker. The one without the Joker has a 'snowflake' back and is a cheating deck where the suit and rank of each card can be determined by small variations in the back design. The deck comes with a card to tell the buyer how the determinations are made.

For a brief period of time, John J. Levy had a partner named Huestis. We have been unable to pinpoint the years that Mr. Huestis was involved. Some believe that it was his capital that financed the building of the original modern plant at 177-179 Grand St. Other people think that Mr. Levy's funds were so depleted after construction, that he was short of working capital and thus had need of a financing partner. We do know that Huestis & Levy organized the Union Card Manufactory, c1855. Huestis & Levy and Union Card Manufactory decks, which all seem to have the same single ended courts, are extremely scarce.

NY17

NY17 JOHN J. LEVY

177 & 179 Grand St., New York, c1860. This is one of the first American decks featuring pictorial backs. The backs portray an unidentified naval battle and the Ace of Spades has the U.S. motto 'E Pluribus Unum' in the ribbon under the eagle. The deck also comes with both single and double ended courts. The beautifully detailed drawing on the back was likely the inspiration for many of the interesting back designs decorating playing cards for years to come.

BROADSIDE FOR HUESTIS & LEVY

NY18

NY18 JNO. J. LEVY
c1865 A different Ace of Spades, also with 'E Pluribus Unum'. This deck also can be found with either one way or two courts. This is the only Levy Ace of Spades with no address.

NY19

NY19 HUESTIS & LEVY
177 & 179 Grand St., New York, c1855. A completely new set of court cards were issued and distributed with the Huestis & Levy (Union Card Manufactory) decks. There were improvements made, as these decks were the first to utilize five color printing. These decks have courts printed in black, blue, red, yellow and green and all known examples are single headed although the pictured advertisement clearly says 'single & double headed cards'. Either there are double headed examples to be found, or Huestis & Levy were also selling the John J. Levy double headed cards.

NY20   NY21

NY20 UNION CARD MANUFACTORY
177 & 179 Grand St., New York, c1855. The identical Ace of Spades and courts as NY19 with but a change of name.

NY21 JNO. J. LEVY, UNION CARD MANUFACTORY
New York, c1855. A new Ace of Spades with the Union Card Manufactory name in the top ribbon and the Levy name at the bottom. It used the NY20 courts.

NY22 JNO. J. LEVY (NYCC) *(not pictured)*
184 William St., c1871. At some point before the merger into NYCC, Levy must have moved into the Lawrence & Cohen premises at the 184 William St. address. This deck has the same Ace as NY16, with only a change of address.

NY22a

NY22a JNO. J. LEVY
123 William St., New York, c1867. The same deck as NY22, but with a different address. It is possible that Levy was at 123 William St in between the move from Grand Street to the Cohen premises. It is known that there was a sales office at that address that was being used in the early 1870s and quite likely it was there earlier.

NY23

NY23 JNO. J. LEVY
184 William St., c1867. A different Ace again with 184 William St. It came with the NY13 Joker also used with NY39.

NY23a JNO. J. LEVY (NYCC) *(not pictured)*
123 William St., c1871. The same Ace of Spades with the 123 William St. address. This deck came with a Best Bower Joker.

Samuel Hart & Co. was, perhaps, the most successful of the successors to L.I. Cohen. Judging by the many examples and variety of his products in collections today, he was the most prolific. He left the Cohen business in 1844, accompanied by another of Cohen's nephews, Isaac Levy (probably a brother of John J.). While preparing to manufacture their own cards, they opened a stationery store at 27 S. Fourth Street in Philadelphia. During this period they called themselves 'manufacturers', but L.I. Cohen produced the majority of the cards they sold.

While slightly outside the realm of playing cards, two old card games found recently, help corroborate the above information. The first, 'Punch's Oracle of Destiny', was published by Samuel Hart & Co. at the above address in 1846. The second, 'What D'Ye Buy?', was published at about the same time by L.I. Cohen at 27 S. Fourth St. Philadelphia.

As discussed in Chapter 3, research has recently demonstrated that Hart and Bartlet were one and the same and that the Crehor Aces were placed on decks made by both Dougherty and Hart. To summarize, evidence clearly points to the following facts. Firstly, Hart purchased the Thomas Crehore playing card plates and some of their machinery. They made Crehor (no 'E') decks, as some of these were definitely made on L.I. Cohen's machine. They also had the typical Hart one way courts. Perhaps, at the outset, Hart wanted to trade on the Crehore name.

Secondly, the Philadelphia directories show that Hart was occupying the premises at 236 S. 13th St. at the same time as Charles Bartlet (one 'T'). Undoubtedly, Hart was Charles Bartlet. Comparison of their first deck (NY24) with the Bartlet deck pictured in Hargrave, p.319, confirms this (refer also to U14 in Chapter 3). Also, Bartlet's later decks had the typical Hart one way courts. Another case of trading on an established name.

In 1858 Hart built a much larger plant at 416 S. 13th Street, where the company remained until the formation of NYCC in 1871. At this time they had the largest sales volume of the three firms and brought with them many patents and innovations. Perhaps it was for these reasons that Samuel Hart was the only name to appear concurrently with NYCC on any newly designed Ace of Spades.

From 1849 on, Hart also maintained sales and business offices in New York City. The following table, from city directories of the period, shows all known addresses in both cities, along with the dates they were listed:

Philadelphia -	27 S. Fourth St.	1844-49
	160 Market St.	1849-51
	17 Merchant St.	1851-53
	236 S. 13th St.	1853-57
	416 S. 13th St.	1858-72
New York -	82 John St.	1849-53
	1 Barclay St.	1854-57
	307 Broadway	1858-59
	560 Broadway	1860-63
	546 Broadway	1864-65
	43 John St.	1866-71

Although readers may not be interested in these addresses, this table helps date Hart cards with reasonable precision. Note that all Hart Aces of Spades with a Philadelphia address were from either 236 or 416 South 13th St.

NY24

NY24 SAMUEL HART & CO.
416 South 13th St. & 307 Broadway, 1858. This deck came with the standard Hart Faro courts as well as unusual one way courts identical to the ones shown on p.319 of Hargrave made by Charles Bartlet. The latter version has square corners and gold edges and features a North American Indian on the backs.

NY25

NY25 SAMUEL HART & CO.
307 Broadway, 1858. This miniature deck (1 1/2 x 2 1/4 inches) is hand colored and stenciled and has European courts patterned after a Wust deck. Perhaps it was imported?

NY26

NY26 SAMUEL HART & CO.
416 South 13th St & 546 Broadway, 1865. This deck is full sized with printed German style courts. The Ace of Spades is printed in red, blue and black. Other examples of this exact deck, but with an anonymous Ace of Spades, have been discovered. It was likely imported as well, with the Ace of Spades being specially printed in Philadelphia for the Hart version.

NY27

NY27 SAMUEL HART & CO.
416 South 13th St & 546 Broadway, 1864. This Ace of Spades was a newer version of NY24, the prototype of the main Ace used throughout the years. The address changed with each move, and slight changes in the print style and flourishes exist. But from this time forward the Ace always had blue print with a black spade pip.

NY29

NY29 SAMUEL HART & CO.
416 South 13th St. & 560 Broadway, c1860. On this Ace of Spades the address is 560 Broadway. This deck has one way courts similar to NY27. It is known with a patriotic theme on the backs, printed in red, white and blue and incorporating the slogan 'Union Forever'.

NY27a

NY27a AMERICAN MANUFACTURE
Phila Card Manufactory, Philadelphia, c1849. A generic Ace of Spades with the familiar Hart Faro courts identical to those of NY27 and NY27b and undoubtedly made by Hart in his early days.

NY29a

NY29a SAMUEL HART & CO.
236 South 13th St. & 1 Barclay St., c1855. Again, the familiar Phila. Card Manufactory Ace of Spades, but this version has the 1 Barclay St. address, making it the first of this type.

NY27b

NY27b HENRY COHEN & CO.
Philadelphia, c1860. Was Henry Cohen another relative, just a name to be used for a different grade of cards or an imitation? The Ace of Spades has similarities to those of some early Cohen and Levy decks, but the courts are identical to those of NY27 and the NY42 Faro decks.

NY30

NY30 SAMUEL HART
1849. The dating of this Ace is clear. Hart are generally thought to have made their first decks in 1849. In addition they operated under the name Samuel Hart from 1845 to 1849 when the name was changed to Samuel Hart & Co. Finally, there is no mention of New York, so the deck must have been made after they started manufacturing in 1849 but before they opened their New York office at 82 John St. later that year. This deck, probably their first, came with one way Faro type courts. It is also known to come with the NY24 courts.

NY28

NY28 SAMUEL HART & CO.
416 South 13th St. & 546 Broadway, c1860. A slightly changed Ace, with the NY address moved from the bottom of the card to the ribbon (replacing 'Phila Card Manufactory').

NY31

NY28a SAMUEL HART & CO. *(not pictured)*
416 South 13th St. & 43 John St., c1860. The same Ace but with the 43 John St. address.

NY31 SAMUEL HART & CO.
416 South 13th St & 560 Broadway, c1860. This was the first of Hart's illuminated decks. It featured a heavily gilded Ace of Spades and newly designed, quite elaborate, courts.

NY31a

NY34

NY31a SAMUEL HART ILLUMINATED
Philadelphia, c1860. This very rare Hart illuminated deck was discovered several years after publication of the original Encyclopedia. It features beautiful, vividly colored one way courts and was likely manufactured about the same time as NY31.

NY34 SAMUEL HART & CO.
(NYCC) 1871. A new multi-colored Ace of Spades was designed to celebrate the opening of the New York Consolidated Card Co. The deck was illuminated with gold on the Ace and used the illuminated courts from NY31. It was listed in their opening price list as #38 Washington (illuminated), the highest priced deck in the line at $168 a gross.

Probably, the most popular and long lived of all Hart issues were the George & Martha Washington Aces of Spades, all published under the Samuel Hart & Co. name. So far, five different designs, published from 1868 until after 1900, have come to light.

NY35

NY35 SAMUEL HART & CO.
416 South 13th St & 43 John St., c1866. Back to the most popular Ace of Spades design, but this time the New York address has been changed to 43 John St.

NY32   *NY33*

NY32 SAMUEL HART & CO.
1866. The first version of the George & Martha deck featured a beautiful, multi-colored Ace, which gave the impression of twin portraits. It also had newly designed, but not elaborate, two-way courts without indices. As it was a high priced, luxury deck, there were nice, multi-colored back designs. This deck was marketed both as a 32 card Euchre deck and as a 52 card deck.

NY33 SAMUEL HART & CO.
1868. As NY32 proved successful, an attempt was made to increase sales by reproducing the pack with the patriotic theme (NY29). Just removing the color from NY32 proved unsatisfactory, so a completely new Ace of Spades was designed. The result was a much younger looking pair. The same double-ended court cards were used although it is also known to have been manufactured with single-ended courts. It also came in a cheaper version still, with a steamboat type back, unusual one way courts and the Imperial Bower Joker pictured under NY36.

NY33a SAMUEL HART & CO. (not pictured)
(NYCC), c1872. The success was surely longlived, as the same Ace of Spades was used, but the address was now 14th Street. Everything else about the deck remained the same.

Samuel Hart produced many decks, often sold in special double boxes complete with rules and scorers, for Euchre and similar games. The following three decks were amongst these and the sets were often found with the great colored Washington Ace (NY32).

NY36

NY36 SAMUEL HART & CO.
Philadelphia & New York, c1863. A newly designed Ace of Spades with a red lion below the Hart name and the initials 'SH' in the spade pip. This Ace was printed in red and black and was made for Euchre. The Imperial Bower was probably the first U.S. Joker!

NY36a

NY36a LONDON CLUB CARD
Samuel Hart & Co., 222-228 West 14th St., New York, c1875. A later version made after the move by NYCC from 184 William St to larger quarters shortly after the merger of the three companies.

NY39   NY39a

NY39 SAMUEL HART & CO.
(NYCC), c1870. All Aces of Spades that merely said Samuel Hart, New York were made by NYCC. The deck also featured the unusual Best Bower with the name 'Consolidated Card Co.' used for NY13. The 'Angel' backs that were a mainstay with Squeezers for many years were introduced with this deck. A separate variation of the Ace is pictured as NY39a.

NY37

NY37 SAMUEL HART & CO.
416 South 13th St & 43 John St., c1868. Finally, a new design for their standard Ace of Spades. The address was 43 John St. and the deck came with a beautiful colored Imperial Bower Joker.

NY38 NYCC *(not pictured)*
(Samuel Hart), New York, c1871. This was the first Ace of Spades to bear the name of the NYCC Co. Manufactory. It was the same pattern as NY37, but the Phila. Card Manufactory was replaced. The pack also introduced the NYCC Best Bower (see NY46).

NY40

NY40 SAMUEL HART & CO.
(NYCC), c1875. The California Poker deck was one of the first produced by NYCC. The Spade Ace showed two 'fortyniners' of the gold rush era and the courts are similar to those of Lawrence & Cohen. The wrapper was marked "For Faro & Poker". This is an extremely rare deck and the only one known is in the USPC museum in Cincinnati. None of the existing NYCC price lists of the era mention the deck and the possibility exists that the deck in Cincinnati was a sample that was never commercially produced.

NY38a NYCC
(Samuel Hart), New York, c1871. The same Ace but with the address placed above the stars.

NY38a

NY41 SAMUEL HART & CO.
(NYCC), 222-228 West 14th St., c1875. The most popular style Hart Aces of Spades, but bearing the new 14th Street address of the New York Consolidated Card Co. This one is quite hard to find, as it was only a short time later that NY42 was introduced, with an Ace of Spades that was to run well into the 20th century.

NY41

NY42 SAMUEL HART & CO.
(NYCC), c1885. This was the most common of all Hart Aces with no indices. It was produced for many years, both as a Poker deck (two way courts) and a Faro deck (one way courts). The later decks that were made for Faro were on lighter stock that had a yellowish tint. This lighter stock also had a glossier finish which made dealing from the dealing boxes easier. This deck has been reproduced in recent years.

NY42

NY42a NEW DEVELOPMENT CO. (not pictured)
1954. A reprint of NY42 that even included an identical reproduction of the box dated 1868. The deck included a Joker with a large eagle and the words 'Joker' and 'Pharo Cards'. It also had an extra joker, stating that it was a deck found in an old gambling house and reproduced. Although the deck was never intended as a counterfeit, some collectors might be confused if the Jokers and box are removed. It is easy to distinguish, however, as the Ace of Spades is printed in black, not the familiar blue, and the stock is dead white, not the buff tone of the original.

NY42b AMERICAN MANUFACTURE
(NYCC - Samuel Hart & Co.), c1890. An 'American Manufacture' deck recently proven to have been made by NYCC (Samuel Hart) and often sold as 'seconds'. Other than the Ace of Spades it was identical to NY42.

NY42b

NY43 NYCC (not pictured)
Samuel Hart, c1900. This was the only no indices deck issued with the NY50 Ace. It became one of the standards for NYCC for many, many years. This deck was issued for export only, as there were a great many overseas customers who would still not accept indices.

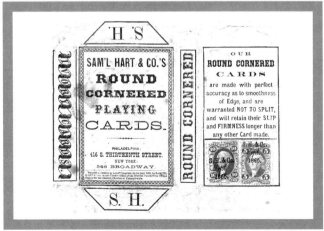

TYPICAL HART WRAPPER

NY44

NY44 SALADEE'S PATENT
Samuel Hart, 1864. Hart likely issued the first deck of American cards with corner indices. It may well be the first deck of indexed cards issued anywhere. The name of the Patent Holder, Saladee, was prominently positioned on the Ace of Spades. The idea, which now seems ridiculously simple, revolutionized the industry. It actually made it possible to more easily play games that required holding large numbers of cards. When one considers that this patent was actually registered in 1864, one wonders why it took over 10 years to gain wide acceptance.

The Samuel Hart Company, as an independent producer of playing cards, existed for a relatively short period of time, from 1849 until 1871 when they joined NYCC. During that 22 year period, they were extremely active and produced a large variety of decks. Every time they moved plants and/or offices they had to redesign a new Spade Ace. During most of that period, single ended and double ended courts were used interchangeably. Styles were changed, and, after their merger, both Levy and Lawrence & Cohen courts were occasionally used.

Before leaving Hart we present an example of the difficulty in dating a deck of cards. A deck and wrapper with the 560 Broadway and the 416 S. 13th St. addresses was found with an 1894 tax stamp attached. Clearly the deck was manufactured in 1860-63. When dating a deck one must remember that tax stamps were required when the cards left the factory, not when they were made! Perhaps this deck was part of a clearout of old decks.

When the New York Consolidated Card Company (NYCC) was formed at the 184 William St. address, it offered only Lawrence & Cohen, Hart, and Levy cards, normally without indices. The NYCC brands were introduced a few years later, likely sometime in 1873, although all three founders names were still being used in letterheads, price lists, etc. as late as 1892.

In the early 1870s the old Aces of Spades were continued, and no mention of the new company could be found, except on the wrappers. Even after their move to larger quarters on West 14th St., probably in 1873, the old names were still being used on some decks. A final note, it is difficult to identify and date the decks in the transition period because inventories were not destroyed when the companies merged and courts from one company might be used with another's Ace of Spades.

The name 'Squeezers' had been used from around 1870 by both Hart and Lawrence & Cohen, but only on their wrappers. The name came from the ability to hold a number of cards in one's hand, and by 'squeezing' them apart, reading their values using only the indices. After 1873, all indexed cards made by NYCC were called Squeezers.

NY45

NY45 AMERICAN PLAYING CARDS, SQUEEZERS #37
SAMUEL HART (NYCC), c1870. A few years after the Saladee patent, a new patented indices deck was released by NYCC under the Hart name. This deck had only two indices instead of the original four. It also had a new Ace of Spades, a modification of NY39 with indices added. The name was indicated only on the wrapper pictured.

NY46

NY46 NYCC
222-228 West 14th Street (office 123 William St.), c1873. This is another of the very early NYCC decks that utilized the Hart patents. This deck was made when the company first occupied the West 14th St. factory. Note the use of the office address at 123 William St. along with that of the factory. The Best Bower (unlike the next listing) did not have the address.

NY46a

NY46a PATENT SQUEEZERS
NYCC, c1873. A newly discovered Ace of Spades, square cornered and similar to NY46 which came with the Best Bower showing the 222-228 West 14th Street address. Perhaps it was the first Squeezer Ace of Spades.

NY46b   NY46c

NY46b SQUEEZERS
New York Card Factory, c1871. Or was this the first Squeezer Ace? Another recently discovered deck which likely is from around the time of the merger. The deck we have does not have a Best Bower.

NY46c AMERICAN MANUFACTURE
New York Card Factory, c1870. The Ace of this deck is virtually the same as NY27a and the indices are very small. The courts, as well as the Ace of Spades, make it likely it was manufactured by Hart just before the merger into NYCC. As the deck has indices and uses the New York Card Factory name, it seems appropriate to list it here.

NY47 PATENT SQUEEZERS *(not pictured)*
NYCC 222-228 West 14th St., New York, c1876. This was the next Squeezer Ace, the same as NY48 except the address is in the bottom ribbon. Note that the Ace and Best Bower (see NY46a) both carry the West 14th St. factory address. This Ace also has the 123 William St. sales office address.

NY47a PATENT SQUEEZERS #19

NYCC, c1876. A beautifully engraved Ace, recently discovered, with George and Martha Washington and the words 'Patent Squeezers #19'. It came in a Steamboat #19 box and included a different Best Bower Joker.

NY48 PATENT SQUEEZERS

NYCC, c1876. We cannot determine the exact date that the complete address was replaced by just 'New York', but it was not too long after the introduction of Squeezers. We do know that the 'Patent' Ace was used until the expiration of the patent in 1890. There were a number of decks of this brand issued to honor celebrities of the day. In 1885 one was made to honor President Grant and in 1887 the back shown, celebrating the Golden Jubilee of the reign of Queen Victoria, was produced. Generals, stage stars, explorers and other famous personalities were likewise honored.

NY49 SQUEEZERS #35

NYCC, c1880. With the expiration of the patent, the Ace of Spades was redesigned. Decks with the pictured aces were sold as 'Squeezers' under #35, #352, #335 Monogram Squeezers and #32 Euchre. At least five versions of the Ace of Spades are known. One, NY49d, has Samuel Hart, along with NYCC, on the Ace of Spades and 'Patent' Squeezers. NY49a shows 'Registered' Squeezers, sometimes with the number 35 and the word 'The' in the middle of the first banner. NY49b has an Ace with less detail and a 'The' added before the name, NY49c has the center Cherub replaced by a wreath and the outer cherubs replaced by the same images used in Gem #53 (refer NY52). NY49, perhaps the first issue, does not have the word 'The' in the name at all. This brand comes with the great Gold Medal Joker shown although, in the early years it is known with a 'Best Bower'.

NY50 HART'S SQUEEZERS

NYCC. This Ace of Spades was by far the most widely used NYCC Ace and three variations are pictured. At least 25 different issues, each with slight variations in the lettering, the leaf arrangement, the dot designs (before the 'C' and after the 'Y') and/or the overall designs in the center are known. The Aces with 'The' are the later issues. These decks first used an early Best Bower but from about 1885-1920 used the more common Best Bower Joker (see NY47a). After that date the 'Squeezer' Joker took over. Besides the Squeezer brand, the NY50 Ace was also used for Anglo American Orientals #910, Bridge Whist #3511, Crow Brand, Crow #177, Dominion #35, Eclipse #432, Full House #1112, Gold Medal #1001, Hart's Crown #444, Hart's Washington #40, Hustler #319, Master #1 Special, Pinochle #97, Pinochle or Skat #298, Ping Pong #888, Pyramid, Skat #29, Sphinx #1144, Steamboats #19 and #220, The Prize #1002, Washington #40 (a private printing for Marshall Fields & Co.), Winner, and, no doubt, many others. Certain brands which used this Ace eventually had their own Aces and Jokers designed. Finally, there was an early version of this Ace made without indices.

NY50a   *NY50b*

NY50a #319 SQUEEZERS STEAMBOATS
NYCC, c1890. Perhaps this was a forerunner to the Hustler #319 brand.

NY50b #2 STEAMBOAT
NYCC, c1910. This deck comes with a generic Ace and the normal Best Bower Joker. The rather plain box has an NYCC tax stamp.

NY50c RUBY QUEEN (not pictured)
NYCC, c1910. This seems to advertise an English cigarette on the backs. It comes in a Ruby Queen box and has both standard and Oriental indices.

EARLY TRITON AD

NY51

NY51 TRITON #42 (TRITON PINOCHLE #997 and #298)
NYCC, c1890. There were at least three different versions of this brand. The first, NY51, used the Best Bower Joker and was issued c1890. NY51a had a similar Ace of Spades, but obviously newer, and came with two types of courts, both different from NY51. It was made c1893 to c1915. NY51b was a later version, probably 1915 to 1925. The same Joker (with some variation in the printing of the name) was used for the two later versions. Triton was a double enameled card on linen stock and one of the more expensive brands offered. It was with this brand that NYCC introduced their new Polychrome finish.

NY52b

NY52 GEM #53 (GEM #57)
NYCC, c1895. These two popular NYCC decks were slightly smaller than the normal size (both in width and length) and were advertised as 'fit more comfortably in the hand'. The brand ran from about 1895-1925. The Ace of Spades (NY52a) was changed once when the 'The' was added (NY52b pictured). Both Pinochle and regular 52 card decks were made with both Aces and the Best Bower, reduced in size, was in all Gem #53 decks. This deck was also issued with point count dots (refer BW13).

NY53

NY53 MONACO
NYCC, 1886, Edgar J. Levey's Patent. This was one of the earliest No-Revoke decks produced. The patent date on the face of the Ace of Spades is March 30, 1886. The entire design of the Ace is reminiscent of Samuel Hart, but the N.Y. Consolidated Card Co. name is on the Ace. The deck had white dots in the center of each major suit (hearts and spades) pip, while the minor suits are normal. The name 'Monaco' was reused for a whist size deck with an Ace similar to NY52 many years later.

The following four decks are listed here, as Gene Hochman apparently had information in the packaging that proved they were made by NYCC. There are also strong similarities to decks made by companies in Chicago that were related to the Longley group and Standard Playing Card Co.

NY54 FALSTAFF *(not pictured)*
Shakespearian Playing Card Co. (NYCC), 1895. A special deck, with a rather plain Ace with the company name and the word 'Falstaff' plus newly designed courts by a company created by NYCC. No effort was made to conceal the NYCC name, as ample information was included within the package.

NY54a RANGER *(not pictured)*
Shakespearian Playing Card Co. (NYCC), 1895. Another deck by the same company. The unusual, handsome courts are the same as NY54/NY54b (shown). There is no name on the Ace and the Joker pictures a steamship (see NY54c). The deck is of Steamboats type quality. The box gives the name of the brand and the company.

NY54b

NY54b OTHELLO
By Shakespearian Playing Card Co. (NYCC), 1895. Another deck in the same series with a more elaborate Ace of Spades and a colorful Joker.

NY54c

NY54c STEAMBOATS #5
Shakespearian Playing Card Co. (NYCC), 1895. A deck with an Ace of Spades like NY54, but with 'Steamboat' replacing 'Falstaff'.

NY55

NY55 MEDIAEVAL PLAYING CARDS
NYCC, 1897. This is the most unusual, and perhaps the most beautiful deck ever produced by NYCC. The newly designed double-ended courts, with minor differences at each end, were dressed in mediaeval attire and printed in six colors. The backs were printed in five colors plus gold. A most unusual box within a box, with the name of the brand printed on the inner portion was used for this deck. As with all newly designed, non-standard court decks, it did not meet with success. It ran for just a short time and very few examples can be found.

NY56

NY56 MASCOTTE #69
NYCC, c1890. This was NYCC's lowest priced 'enameled' cards. The first issues had the Best Bower, but later had their own 'Ye Joker'. In some versions 'Ye Joker' was multi-colored, in others he was more subdued. These decks came with the same three sets of courts as Triton, NY51.

We continue with the history of the New York Consolidated Card Co. A meeting was held on July 20th, 1894 between NYCC and the United States Playing Card Co. to discuss the purchase of the assets of the Perfection Playing Card Co. of New York. It was decided that USPC would purchase Perfection and shortly after, NYCC offered to sell their company to USPC. A deal was made and in 1894 NYCC became a part of the United States Playing Card Co., although separately operated. As part of the deal, Messrs. Lawrence and Cohen were elected to the USPC Board of Directors.

By 1877 the NYCC 'Squeezers' were a great success and they were continually expanding to meet the demand for these indexed decks. Court rulings had decreed that any cards made with 'numerals' in the corners infringed on their patents. Their only competition came from the Andrew Dougherty Co. (Chapter 6), who were making their own patented 'Triplicates'. This brand had miniature cards in the corners instead of numbers, but could still be 'squeezed'. In an historic pact, the two companies divided up the country as a means of eliminating outside competition.

To honor this agreement, NYCC issued a 'Squeezer' deck with an unusual back design. Two dogs, Squeezer and Trip, were chained to their respective doghouses, with a quote at the bottom reading 'There is a tie that binds us to our homes'.

In the early 1900s, NYCC moved its main plant to much larger quarters at 4th & Webster Aves., Long Island City, N.Y. No positive information exists as to the date that NYCC closed the 14th Street plant. We know that it was still in operation in 1900 as NY59, Century #191, which was issued to commemorate the turn of the century, had the 14th Street address on the Ace of Spades. Also, a wholesale price list from NYCC dated 1913 still gives this address although most, if not all manufacturing was taking place at the Long Island facility by that time.

From 1894 until 1930, NYCC operated as a separate company, under the umbrella and guidance of USPC. On June 4th, 1930 the New York Consolidated Card Co. and the Andrew Dougherty Co. (which became part of USPC in 1907) merged into Consolidated-Dougherty Co., with headquarters in New York. On September 1st, 1962 this company was dissolved, and thereafter operated as a division of the United States Playing Card Co. of Cincinnati. Interestingly, a few of the original NYCC brands (e.g. Bee #92) are still being made today.

NY58

NY58 ORIOLE #912
NYCC, c1915. This deck features a beautiful multi-colored Ace of Spades and Joker. The birds and foliage are in full, accurate colour.

NY57

NY57 SQUEEZERS
1877. The original deck had the NY48 Ace of Spades. The same back was continued as part of the Squeezer line until about 1915, and it was reissued briefly in 1927 (50th anniversary) and 1977 (100th anniversary). Most of the decks found today have the NY50 Ace.

NY59

NY59 CENTURY #191
NYCC, 1900. This deck was issued to celebrate the turn of the century. It was only issued for a short period and came with the Best Bower Joker.

Some of the brands included under the general heading 'Squeezers' are Bee #92, Elf #93, Gem #53, Hart's Crown #444, Hart's French Whist #68, Hustler #94, Monaco #1909, Mascotte #69, Steamboats #220 and Triton #42 together with variations of these for Solo, Euchre, Pinochle, Skat, etc.

NY60

NY62a

NY60 BEE #92

NYCC, c1895. It is still being marketed today, making it one of the longest running brands made anywhere. The design has changed very little through the years. The first Ace of Spades had intricate detailing in the design, which has gradually been simplified. Conversely, the black and white Joker, which was simple, has gradually added details and is now in full color. NY60 was made until about 1910, NY60a until about 1925, and since then there have been minor variations leading to the current version with the name 'Consolidated-Dougherty' on the Ace of Spades, but USPC on the Joker. Finally, mention should be made of the Centennial Edition of Bee #92 in 1992, which was issued in limited quantities in both the U.S. and Canada to commemorate the 100th anniversary of this brand.

From here to the end of the section, we will attempt to show all of the other brands known to have been made by The New York Consolidated Card Co. Many previously unlisted brands have been discovered, and others, previously listed in miscellaneous categories, are now known to have been made by NYCC. In order to retain the numbers assigned in the original Encyclopedia, the decks are not listed in chronological order.

NY62a DELUXE #142

NYCC, c1921. This bridge size deck came with a multi-colored Ace of Spades and Joker together with vividly colored 'poly-chrome' floral design backs. It was the most popular deck of this size made by NYCC. The Deluxe brand was also used to market NYCC's narrow named cards, which competed with the narrow Congress cards, issued at that time by USPC.

NY63

NY63 ELF #93

NYCC, c1890. This was the only miniature brand ever sold by NYCC. It had the unusual dimensions of 2 15/16 x 1 15/16 inches. Their advertisements read 'Adapted for play where space is limited, but large enough for any game.' NY63 had a beautifully engraved Ace of Spades. NY63a was the same design but cruder. NY63b, introduced in 1910, was based on the Samuel Hart Ace (NY50). Both decks had a standard Best Bower.

NY61

NY61 AUTOMOBILE #192

NYCC, c1901. Perhaps made to complement the Bicycle brand, this deck has a special colored Ace and automobile Joker although the standard Best Bower is frequently found. There were several back designs, all showing people in early cars.

NY62 DELUXE #342 (not pictured)

NYCC, c1920. A very attractive wide deck, with a multi-colored Ace of Spades and Joker offered for several years but never very popular. The Ace and Joker are like the narrow version issued shortly after (NY62a).

NY64

NY64 HART'S FRENCH WHIST #96

NYCC, c1905. One of the earliest of the whist size decks by NYCC with a great colored Joker. The back designs incorporated the fleur de lis and minor variations in the Ace have been found.

NY65

NY65 BEE FRENCH WHIST #68
NYCC, c1910. Another early whist size deck, made for only a short period.

NY65a

NY65a BEE BRIDGE WHIST
NYCC, c1920. The identical deck as NY65, but the word 'Bridge' was substituted for the word 'French'. The same #68 was used, and the Joker was identical.

NY66

NY66b

NY66 FREE LANCE #915
NYCC, c1920. This brand in both wide and narrow (NY66a) versions used either the pictured special Joker or the Best Bower.

NY66b FREE LANCE #915
NYCC, c1925. The identical number was used for a new version of this deck, made in Bridge size only. It had a newly designed, multi-color Ace of Spades and Joker and was colorful and beautiful.

NY67

NY67 BEE BRIDGE
NYCC, c1930. Another Bridge/Whist size deck capitalizing on the popularity of the 'Bee' name and advertising. The colorful Flapper Joker is only found with this deck.

NY68

NY68 STERLING WHIST #196
NYCC, 1920. The design of the Ace of Spades and Joker were borrowed from NYCC's most beautiful deck, Mediaeval (NY55). The deck was long and narrow (2 1/4 X 3 3/4 inches) with standard courts. It was a top-of-the-line deck, packaged in a slip case.

NY69

NY69 NYCC
c1910. Once again, the popular Hart Ace was used. This is an example of the narrow version of NY50. This type of Ace was used when manufacturing private brands for department stores, drug chains, etc., using the customer's name and designated brand on the box, but with no special Ace of Spades or Joker. With this Ace, the Best Bower was usually used. It is also known with a floral design back and the NY62a (Deluxe) Joker.

NY70 DRUMMER *(not pictured)*
NYCC, c1890. This was a Steamboat grade deck made on cheap stock. The Ace of Spades is very similar to Samuel Hart's (NY39). The Best Bower was used.

NY71   NY72

NY76

NY71 HUSTLER #94
NYCC, c1890. A scarce brand which likely came with the Best Bower Joker.

NY72 UWINNA
NYCC, 1915. The Ace and Box featured a Swastika (Indian Good Luck sign). It also came with the Best Bower.

NY76 HART'S CROWN #4444
NYCC, c1910. Another deck listed in the export brand section (note the four corner indices). The box has the Long Island City address of NYCC.

NY73

NY73 A1 THE BEST
NYCC, c1905. No maker's name appeared on the Ace of Spades or Best Bower Joker. It has identical courts to those of other NYCC brands of the 1900 era.

NY77

NY74 FAN-TAN (not pictured)
NYCC, c1910. A plain Ace of Spades designed similar to A-1 The Best.

NY77 NESTOR
(formerly MSW110), Popular Playing Card Co. (NYCC), c1910. Another cheaper generic brand with a great Joker known to have been made by NYCC.

NY75

NY78   NY79

NY75 CARLTON #919
NYCC, c1910. A high quality deck with the Ace of Spades and Joker printed in black and red. It was listed in their 1910 price list as Carlton #919 for export, thereby explaining why it had four corner indices. The deck shown was made for a customer in Alexandria, Egypt.

NY75a CARLTON #919 (not pictured)
NYCC, c1910. This deck was made as a private brand for domestic use (two corner indices). The similar Ace of Spades was in black only and it came with the Best Bower. The NYCC name does not appear on the cards or the box.

NY78 CANARY #911
(formerly MSW97), c1910. This brand used the Best Bower Joker and came with a variety of monotone backs.

NY79 MONTE CARLO #11
NYCC, c1905. The Ace of Spades is similar to NY52 (Gem) with the wording changed and the Mascotte Joker (NY56) was used. It was sold as 'French Size'.

NY79a MONACO #1909 (not pictured)
NYCC, c1909. The same deck with a different name printed on the Ace.

NY80

NY80 MONOGRAM SQUEEZERS #335
NYCC, c1920. This was the same fine quality card as their regular Squeezer #35 brand. Their sales of the Squeezer brand especially made for Clubs, had increased to such an extent that they were now offered with monogramed Aces of Spades, Jokers and backs.

NY82a

NY82a BAILEY, BANKS & BIDDLE
c1915. Another deck made by NYCC for them used at least two different Jokers and came with both NYCC and USPC courts.

NY81

NY81 LIGHTHOUSE #922
(formerly MSW95), c1920. A deck, with a great, colored Joker, now identified as having been made by NYCC.

NY81a LIGHTHOUSE #927 PINOCHLE *(not pictured)*
(formerly MSW96), c1930. The Pinochle version of NY81 with the New York Consolidated name appearing only on the box.

NY82b

NY82b BAILEY BANKS & BIDDLE
(formerly MSN10), c1915. A 'Playwell' deck manufactured by NYCC.

NY83

NY83 GEORGE B. HURD
c1920. This company, perhaps better known for its no-revoke deck (NR10), was actually a stationer in New York, not a maker of playing cards. It has now been established that George B. Hurd's cards were made by NYCC. This deck has a beautiful, colored Ace of Spades and Joker.

NY82

NY82 BAILEY BANKS & BIDDLE
(formerly MSN9), c1915. This deck was made for the large Philadelphia Jeweller and Stationer. Several manufacturers made special decks for this company.

NY83a

NY83a GEORGE B. HURD
(formerly MSN38), c1920. A slightly less vibrant version came with a similar Joker.

NY86

NY86 FRANCO-AMERICAN #112
NYCC, c1900. This version of #112 has the NYCC name on the courts. Two other versions are known, both without the name on the courts. One has four corner indices and one has none. These three decks all had a plain pip for the Ace of Spades and some came with the Mascotte Joker.

The New York Consolidated Card Company produced several decks with Spanish suits. On one of these, the identity of the maker is actually known, on another it can be surmised to have been made in Barcelona, and on the third it is assumed to be made in the United States by NYCC or USPC.

NY87

NY87 NAIPES FINOS #28
NYCC, c1895. This deck very closely resembles those made in Barcelona at the end of the 19th century. It was only made for a short period by NYCC.

NY84

NY84 SQUEEZERS
NYCC, 1878. This deck with Spanish suit signs, was originally made to be sold at the Paris Exposition in 1878. NYCC continued making it into the early 20th century.

Samuel Hart also made gaming tokens in white metal, brass, copper and hard rubber. Sometimes the copper and brass tokens were silver plated. The large tokens were made in the 1854-57 period and the small ones in 1858-59. The hard rubber ones (extremely scarce) seem to have been made during the Civil War and certainly were made in the 1860-64 period. Incidently, the large token had a 'typo' indicating the Philadelphia address as '36th St.' not '13th St.'.

NY85

NY85 SPANISH
Samuel Hart, 1850. This deck, which bears the date 1850, was made by Real Fabrica de Madrid, probably for Samuel Hart. The only evidence of Hart was on the wrapper, and the initials 'S.H.' on the 4 of Cups. It also had a red, white and blue bunting on the Ace of Swords and a red, white and blue motif in each Coin.

1854 TOKEN

ANDREW DOUGHERTY

Andrew Dougherty's story is a key part of the development of the American playing card industry.

Andrew Dougherty was born in Donegal in Northern Ireland in 1827. In 1834 his parents came to the United States with young Andrew and settled in Brooklyn. His first taste of the playing card business was at a very tender age when he was employed by David Felt, a stationer and early New York maker of playing cards (refer Chapter 3). Dougherty had a varied career, trying many occupations including a two and a half year trip as a seaman on a whaling ship out of New Bedford, Mass., before re-entering the playing card field with an unknown maker, possibly Crehore, in 1844. By 1848 he had become a foreman and had saved the rather princely sum of $800.

With these savings he decided to establish a playing card business in New York, the city incidentally, which was always the home of the Dougherty playing card business. The records indicate that he started his business in a loft in Brooklyn at 48 Ann St. in 1848. With the assistance of a boy and a girl he produced two gross of decks a week. While no one has been able to identify any playing cards issued under his name at that address, there is no doubt that the early Dougherty cards without a street address were made during this period.

In late 1849 he moved to 78 Cliff Street in New York City. In need of capital, he entered into a partnership with two brothers named Coughtry. Coughtry and Dougherty continued in business for three to four years, but not very successfully, and by 1853 the partnership had dissolved in debt and Dougherty was once again on his own, having paid off all his debts. He was fortunate, at that time, to be able to make an arrangement with a large purchaser of his cards to furnish capital in exchange for the equivalent in manufactured goods. This arrangement remained in force over the next seven years and his business improved greatly.

By 1858 Dougherty was quite prosperous and moved his rapidly expanding business to five floors at 26 Beekman Street connecting with four floors at 28 Beekman Street. He prospered there for another 15 years and in 1872, to again increase his manufacturing capability, he moved to the northeast corner of Centre and Worth Streets where he completely remodeled and rebuilt the existing building.

When completed, the seven story main building at 76 Center Street connected with a smaller, fireproof building used for storing plates, dies and rollers and containing a complete machine shop. The basement contained a 40 horsepower steam engine to drive the machinery and two steam boilers to furnish heat for

warming the building and drying the cards. All the floors were lit by gas, which was necessary only in the winter months since they were well lit by the many windows. Water was supplied on all floors and there was even an elevator.

When Dougherty began making playing cards with two employees in 1848, all cards were made by hand, with the colors being added to the printed (from wood blocks) courts by brush and stencil. By 1870, his processes were automated and his machinery, most of which he had invented and developed himself, cost over $70,000 and he employed 130 persons. One machine printed twenty complete packs per minute and the total capacity of his establishment was three and a half million packs a year. By the early 1870s he had developed a machine to produce cards with accurately rounded corners and all his cards were made that way, except those sold to gambling casinos which required square corners to fit their card racks and dealing boxes.

Dougherty can be credited with several innovations to the cards themselves. For example, he was the first United States maker to introduce the 'Little Joker', which was derived from a European model. Also, in

1876, he secured a patent for Triplicate Cards, with a miniature facsimile of the face of the card in opposite corners, as a competitive response to the Squeezers manufactured by Hart and the other NYCC constituents. Again, in 1883 Dougherty secured a patent for a corner index (which varied slightly from the NYCC version) under the name Indicator cards. These indices were at first added to the Triplicates, but in 1884 the Triplicates were discontinued and Indicators became the norm.

Dougherty made playing cards of many grades of quality, from the finest for private clubs to the cheapest for the low end retail market. According to a journal article in 1876, Dougherty had already used at least 260 special back designs over the years, some of which were copies of those used by Jazaniah Ford, Owen Jones and other earlier makers.

He also made cards with Spanish suit signs for sale in California, Texas, Mexico, South America and even in Europe beginning in 1882. In the early 1900s Dougherty manufactured Austro-Hungarian cards in the Seasons pattern. His cards were elegant, durable and slipped easily through the hands because of the two patented finishes he developed and used for them. The first was called Linoid and the second Pegulose.

By 1896, when Dougherty turned the business over to his three sons, he had made it into one of the leading manufacturers in this field, not only in the United States, but also in the world. He died in 1905 at the age of 78 and his sons continued to operate the company in New York until 1907 when the United States Playing Card Company purchased it. USPC kept the Dougherty business operating independently until 1930 when it was combined with the New York Consolidated Company, long a subsidiary of USPC, to form Consolidated-Dougherty, a division of the United States Playing Card Co.

AD1

AD1 ANDREW DOUGHERTY
78 Cliff St., c1850. An early deck of Cliff St. cards with strange, and somewhat homely courts with a somewhat European flavor. It is recorded that Dougherty manufactured Spanish cards dated 1849 while at this address, but we have been unable to confirm this by locating actual cards.

AD1a

AD1a AMERICAN CARDS
Andrew Dougherty, 48 Ann St., c1848. This is one of his earliest, probably the first, deck as no address appears on the Ace of Spades. The courts are much more handsome than AD1 and, of special interest, are identical to those of Thomas Crehor U5. It is interesting to note that Crehore's business was continued until 1846, almost the same time as Dougherty started, and we have assumed that this deck was made at 48 Ann St. in his first year of manufacture.

AD1b ANDREW DOUGHERTY *(not pictured)*
78 Cliff St., c1855. This Ace of Spades is identical to AD1, but the courts are the double ended type used by Dougherty until the advent of his 'Triplicate' and 'Indicator' decks.

AD1c

AD1c ANDREW DOUGHERTY
78 Cliff St., c1850. Again the same Ace of Spades as AD1, but this time with one way courts like those of Thomas Crehore U4. As discussed in Chapter 3, these courts were used by Crehore, Crehor, Bartlet and Hart, as well as Dougherty, and were the courts that continued into the 1900s in the Hart Faro decks made by NYCC.

AD1d ANDREW DOUGHERTY
26 Beekman St., c1858. Another Cliff Street style Ace of Spades, but this time with the new Beekman St. address. Clearly this was the first Dougherty deck made at the new factory. The courts are again similar to those used in Thomas Crehore (U4) decks.

AD1d

Readers are perhaps wondering about the cards made by the partnership of Coughtry & Dougherty. We can presume that very few were made under that name, as we have only seen two examples. Probably, decks with just Andrew Dougherty's name were also being manufactured during the few years of the short-lived partnership. The Coughtry & Dougherty examples have been moved from the Early Makers category as it has been clearly established that they were manufactured by Dougherty.

AD1e

AD1e U.S. CARD MANUFACTORY
(formerly U16), Coughtry & Dougherty, New York, c1851. This deck has a very colorful Ace of Spades and is illuminated, which means that the pips are outlined in gold leaf with the courts making extensive use of the gold leaf as well. The courts are identical to those of AD3.

AD1f

AD1f COUGHTRY & DOUGHERTY
New York, c1851. Another deck with the partnership name and a different Ace of Spades, printed in only one color. It is interesting to note that the eagle faces in a different direction and there is no American flag above it.

AD2 EXCELSIOR
26 Beekman St., c1859. As previously stated, it is impossible to determine an exact date for any deck found without a dated tax stamp, dated wrapper or some other persuasive evidence. The Beekman St. Aces with the shaded center wreath are known to be earlier than the ones without shading. Except for AD6, the Beekman Aces come with decks with both one way and two way courts, Poker players preferring the two way courts with Faro players demanding the one way variety. As a final comment, almost all of the no indices Dougherty Aces are labeled 'Excelsior'.

AD2

AD2a EXCELSIOR *(not pictured)*
26 Beekman St., c1861. The same Ace with two way courts.

AD3

AD3 ILLUMINATED
26 Beekman St., c1865. This is a beautifully illuminated deck with courts identical to those issued by the U.S. Card Manufactory (Coughtry & Dougherty). This deck came in a colorful box entitled Great Mogul Playing Cards.

AD4   AD4a

AD4 EXCELSIOR
26 Beekman St., c1864. A second Beekman St. Ace with the shading removed from the inner wreath. The printing of the Ace of Spades is in blue.

AD4a EXCELSIOR
26 Beekman St., c1864. This deck, with a marked back was like AD4, but the name and address were missing. It appears that marked decks were made by many of the early manufacturers but their names were usually omitted as they did not want to be identified with cards that could be used for cheating.

AD5

AD5 EXCELSIOR
26 Beekman St., c1864. This Ace is slightly different than AD4. The deck is of better quality, and again the Ace is blue. AD4 and AD5 come frequently in shorter decks (32 or 36 cards) for Solo, Bezique and Euchre in addition to the regular 52 card decks.

AD6 EXCELSIOR

c1870. The was the last Ace of Spades to be introduced at the Beekman St. address. This deck was the first to be made with round corners. It was also the first to be made only with two-way courts.

AD6

After the move to 76, 78, & 80 Centre St., most of the Beekman St. Aces of Spades were continued, with only a change of address. Innovations and improvements came regularly with round corners replacing square, Best Bowers and then Jokers being added and indices, starting with the Triplicates, becoming the standard. Dougherty, about this time started making special advertising decks, foreign suited decks, and other decks of speciality manufacture. Interestingly though, they never made miniature or patience size decks.

AD7

AD7 EXCELSIOR

Andrew Dougherty, 78 Centre St., c1872. This was the longest running Ace with the Excelsior name. It was introduced at Beekman St. in the illuminated deck (AD3), and continued to the end of the 'no index' era. It was with later issues of this deck that Dougherty introduced the 'Little Joker'. The deck came with both two way and one way courts and a King of Hearts from each is shown. Earlier versions of this listing came with a blank card or the AD8 Best Bower. A final point of interest, many of the early manufacturers designed backs which advertised their company, like the back shown.

AD8

AD8 GREAT MOGUL

Andrew Dougherty, 78 Centre St., c1872. This was the same deck as AD6, with the change of address to Centre St. The Best Bower was also used with this version.

By this time several brands had been introduced by Dougherty, including Excelsior, Mogul, Mogul #15, Henry VIII, Barcelona #49, Indicator #50, etc.

WRAPPER

AD9   AD10

AD9 EXCELSIOR

Andrew Dougherty, 78 Centre St., c1875. A slightly later version of AD8. Note the A. Dougherty name in the center of the card was changed to be more readable. The courts were the same as in AD6/AD8. The deck pictured here had a 34 star U.S. flag back, which helps to date it. The deck was produced with both a blue and a black Ace of Spades.

AD10 EXCELSIOR

Andrew Dougherty, 78 Centre St., c1872. This deck is the Centre St. variety of AD4. The deck was the same in every respect, except for the address. The design of the double ended courts changed very slightly through the years and the changes were not deliberate, rather the result of changing the plates as they wore out.

On September 12, 1876, Andrew Dougherty secured his first patent for Triplicates, a novel type of indices with a miniature card in the top left and bottom right corners, making three images in all. These served the same purpose as NYCC's Squeezers, and while this innovation lasted but a few years, it helped Dougherty become a major force in the playing card industry. While the fancier Ace was thought to be the first, recent information, including the revelation that the Centennial deck of 1876 (SX2), has the plainer Triplicate Ace, has proved that AD12 was the earlier of the two. It should be added that there are many variations in the size of the Triplicate indices, with the smallest ones being the earliest.

AD12a   AD12b

AD12a TRIPLICATE #18
Andrew Dougherty, 78 Centre St., 1883. The last issue of Triplicates had the indicators of AD13 added.

AD12b A. DOUGHERTY
New York, c1873. This previously unknown Ace of Spades is likely the forerunner of the Triplicate Ace. The deck has square corners, no indices and the Best Bower. Note the eagle, bison, U.S. flag and deer that decorate the Ace.

AD11

AD11 TRIPLICATE #18
Andrew Dougherty, 78 Centre St., 1878. This unforgettable deck had an intricate and beautifully engraved Ace of Spades. Note the cherubs at the bottom and the Old English style printing at top, with the spread of cards demonstrating the innovation inside the Ace of Spades. The Little Joker (see AD7) was used.

AD13

AD13 TRIPLICATE
Andrew Dougherty, 78 Centre St., 1883. On May 29, 1883, a new patent was granted to Dougherty for Indicator cards. In effect, it was the same as NYCC's Squeezers, with just enough minor changes to allow for a new patent. For a short period, the printed Indicator was added to the popular Triplicates. This new Ace of Spades carried the dates of both patents and contained the Jolly Joker. The suits are distinguished in the Triplicate indices by the Spanish method of margin spacing.

While the Indicators patented in 1883 soon became the standard for use in all Dougherty decks, there must have been some card players who preferred Triplicates because they were still listed as an available brand in the July 1891 price list.

AD12

AD12 TRIPLICATE #18
Andrew Dougherty, 78 Centre St., 1876. The Ace of Spades was a far simpler pattern. There were no cherubs at the bottom nor eagle at the top, the printing was simpler and the quality of the engraving decidedly less. The early decks contain the Best Bower with subsequent ones containing The Little Joker and, sometimes, the new Jolly Joker which became the standard for most of the decks of the Indicator brand.

AD14

AD14 AMERICAN PLAYING CARDS
A. Dougherty, c1878. One of the last no indices decks to be issued by Dougherty. This Ace was the basis for all those of the Indicator series.

AD14a AMERICAN PLAYING CARDS *(not pictured)*
A. Dougherty, c1883. The same Ace with the first Indicator indices added. Both these decks used the Jolly Joker.

AD15

AD15 INDICATOR

A. Dougherty, 1884. This brand proved to be one of Dougherty's biggest sellers for over 35 years. With the introduction of Whist size cards, Indicator was even manufactured in the narrower width. At that time several other brands were also created using this Ace, with merely a change of name.

AD15a   AD15b

AD15a INDICATOR #50

A. Dougherty, c1896. The number '50' was added at the bottom of the Ace of Spades.

AD15b INDICATOR #14

A. Dougherty, c1895. The number '14' was added to the bottom of the Ace of Spades. It is not known which of the numbers came first, but it seems logical that it was #14. Unlike the #50, this deck is seldom found.

AD15c INDICATOR #14 (not pictured)

A. Dougherty, c1895. A Whist size deck, with a narrow Jolly Joker, was produced for a very short period as new whist and bridge decks were soon developed to meet the competition from NYCC and USPC for this style of cards.

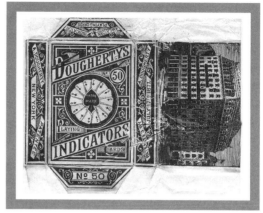

EARLY INDICATOR WRAPPER

Many brands utilizing the Indicator Ace of Spades were issued, e.g. Black Cat #50, Climax #14 (see also AD37), Eden #1899, Euchre #28, Ivorette #70, Lino #1407, Solo #36, Skat #38, Sunset #12, Mogul #15 (AD14 Ace), Oriental #91, Tandem #84 and Tudor #1485. The only distinguishing feature of these decks was a different box and, accordingly, they are not listed as separate entries. Several different Jokers, including the familiar Jolly Joker and the ones pictured, were used.

By this time Dougherty was also producing a variety of advertising decks. We have seen such decks with AD15 Aces of Spades and others as early as AD8. Dougherty also manufactured a number of advertising decks with special Aces and Jokers, although they were not as prolific in this regard as USPC.

From the 1890s on, Dougherty used either Andrew Dougherty or A. Dougherty on his Aces of Spades. As a general rule, the longer version, which often included the word 'Manufacturer', was used on the earlier versions of each brand.

AD16

AD16 JEWEL #95

A. Dougherty, c1900. The same Ace was used for a smaller size deck, advertised as "fits more comfortably in the hand". The cards were 2 3/6 x 3 3/16 inches. The deck had its own Joker, a Queen with a diamond tiara.

AD18 BRIDGE WHIST #45
A. Dougherty, c1900. This brand was developed as Bridge Whist, a transitional game to Auction Bridge. It used both the Jolly Joker and the Clown Head Jokers of AD15.

AD18

AD16a JEWEL #95
A. Dougherty, c1905. This brand must have been successful, as it was soon given a newly designed Ace with a different jewel Joker. It still carried #95 (#95x with gold edges).

AD19 BRIDGE #45
A. Dougherty, c1910. Bridge Whist did not last very long and another similar brand was produced for Auction Bridge players, again with a Jolly Joker.

AD19

AD16b POLO #80
A. Dougherty, c1900. Another deck was issued in the same smaller size as Jewel, with a Polo Ace, its own Joker and a colorful wrapper inside a nice slipcase. Polo #80 was also issued in a tuck box with the Jester on a stick Joker of AD15.

AD20 BRIDGE #345
A. Dougherty, c1910. The last of the Indicator Aces, this time with a multi-colored Ace of Spades and Jolly Joker.

AD20

AD17   AD17a

AD17 TOURNAMENT WHIST #63
A. Dougherty, c1900. A newly named Whist size deck with an Indicator style Ace. It had a narrow version of the 'Climax' Joker (see AD37).

AD17a TOURNAMENT WHIST #63
A. Dougherty, c1900. A similar version, probably earlier than AD17.

AD21 PATENTED
A. Dougherty, c1882. The Patented Ace of Spades was the last of the no indices decks issued. Presumably 'Patented' referred to the indices, so it is interesting that this Ace was used for a deck without the indices. The same Ace of Spades and Joker (girl on stick - see AD15) were used for a Double Pinochle #78 deck which came in a special box with double tax stamps dated 1906.

AD21

AD21a

AD21a PATENTED
A. Dougherty, c1885. The same Ace, but with indices. AD21 and AD21a were issued in both square cornered and round cornered versions and used for many different brands. The Jolly Joker was also used with a scarce brand, American Club #60.

AD21b

AD21b SPANISH-AMERICAN
Andrew Dougherty, c1883. A rare deck made for the Spanish community in the U.S. but utilizing the Patented Ace of Spades and designed to make it easier for those accustomed to Spanish suits to adapt to the popular games in America. As well as the regular indices, the numbers 11 to 13 were used on all courts. Separations at the borders, always present on Spanish suited decks (e.g. AD13) were also included. The Ace and Best Bower Joker were printed in blue.

AD22

AD22 TALLY-HO #9
A. Dougherty, 1885. This famous brand was introduced in 1885 and is still available today. The pictured earliest Ace had the Centre St. address and the Jolly Joker was used until the well-known TallyHo Joker was introduced in the early 1900s. The brand has had very few changes throughout the years although minor variations were made over the years.

STORE CARD

AD23

AD23 STEAMBOAT #0
Andrew Dougherty, c1888. Like every other manufacturer of the period, Dougherty's Steamboats were the cheapest cards in their line. This 'Andrew Dougherty' Ace of Spades was issued in about 1888. A similar Ace with the word 'patented' replacing 'No. 0 Steamboat' came in a box with the name 'Indicators No. 0'.

AD24

AD24 STEAMBOAT #0
A. Dougherty, c1900. The deck was identical to the above except for the modifications to the Ace of Spades. The pictured Joker was used only with Steamboat #0.

AD25

AD25 NAIPES FINOS #49
Andrew Dougherty, 78 Centre St., 1882. Records show that Dougherty manufactured Spanish suited playing cards over a large span of years. It is believed that they were made for shipment to Texas, Mexico and Latin America. The 4 of Coins was always used by the makers for identification of Spanish decks. Another custom has the 5 of Batons carrying the date of manufacture. As shown on the 2 and 5 of Coins, Dougherty was proud of the awards won at the Centennial Exposition of 1876 and the Paris Exposition of 1878.

AD26 NAIPES SUPERIORES *(not pictured)*
A. Dougherty, 1882. A second, quite similar deck and although the same year is printed on the 5 of Clubs, it is doubtful that this deck is as early as AD25. Examples of this deck still carrying the 1882 date are known with tax stamps as late as 1915. This deck, (not pictured) just says 'A Dougherty, Naipes Superiores, New York' on the four of Coins.

After almost half a century in the playing card industry, Andrew Dougherty transferred control of the business to his three sons in 1896. The firm continued to prosper under the leadership of the Dougherty sons and during the next 11 years introduced a variety of new brands. However, after the death of their father in 1907, Edward J. Dougherty, Andrew Dougherty Jr. and William H. Dougherty agreed to sell the company (through a complicated stock transfer) to the United States Playing Card Co.

USPC operated the company as a separate entity until June 4th, 1930, when it was merged with NYCC to form the Consolidated-Dougherty Company, as discussed in the NYCC chapter.

The great majority of the decks shown in the balance of this chapter were made after 1900, many after the purchase by USPC. Occasionally, USPC courts were used in a Dougherty deck, and many of the progressive ideas of both firms were interchanged. Also, manufacturing for brands of both companies was carried out in both New York and Cincinnati.

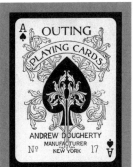

AD29

AD29 OUTING #17
Andrew Dougherty, c1892. The first issue of Outing introduced a popular and long running brand (to about 1925), although this Ace of Spades was made for only a relatively short time. The brand came with a Jolly Joker (see AD12) initially but soon switched to the successful Fisherman Joker (sometimes without the word Joker in the corners) for the remaining life of the brand. The backs always featured game fish, animals or birds and it was manufactured to compete with competitors' brands such as Sportsman and Sporting.

AD27

AD27 KLONDIKE #4711
A. Dougherty, c1910. An interesting deck, undoubtedly issued to commemorate the gold strike in the Klondike. The deck is very rare and the Joker features a prospector panning for gold.

AD30

AD30 OUTING #17
Andrew Dougherty, c1895. A completely new Ace of Spades, but still with the Andrew Dougherty name. Note the snipe in the pip, and the moose and deer on either side.

AD31

AD31 OUTING #17
A. Dougherty, c1905. The last Ace of Spades.

AD28

AD28 FAIR PLAY
A. Dougherty, c1910. Another rare brand with a clown juggling six balls as the Joker.

AD32 RED SEAL #16 *(not pictured)*
A. Dougherty, c1915. There are many varieties of this brand, but they are quite rare. This is the wide, poker size deck, identical in every way (except width) to AD32a. The same Joker was likely used for each version which required one.

AD32a

AD33

AD32a RED SEAL PIQUET #161
A. Dougherty, c1915. This was a Whist size version of the same Ace, issued in a 32 card deck (7 to Ace) for Piquet.

AD32b RED SEAL #16 (not pictured)
A. Dougherty, c1915. A 53 card Whist size deck.

AD33 RED SEAL PINOCHLE #016
A. Dougherty, c1910. This version likely was the first issue of the Red Seal decks. The deck actually had a red seal in the center of the Ace. This Ace came in a standard 52 card deck, a Pinochle deck and a 52 card Whist deck.

AD34

AD34a

AD34 CRUISER #96
Andrew Dougherty, c1897. This deck was dedicated to the U.S. Navy by the company. The Ace of Spades has a warship, and the Joker is a sailor walking with a goat (the mascot of the Naval Academy at Annapolis).

AD34a CRUISER #96
A. Dougherty, c1905. The later A. Dougherty version, which also came in a Battleship box, has an Ace with the simpler name.

AD35

AD35a

AD35 FINISHED WITH PEGULOSE
Andrew Dougherty, c1905. This was a new substance designed to give an extremely glossy finish to the deck to facilitate dealing and shuffling. It was used on many of their better brands and in these cases was noted only on the flaps of the boxes.

AD35a FINISHED WITH PEGULOSE
Andrew Dougherty, c1910. This came in a deck in a Tally-Ho box with a Tally-Ho Joker.

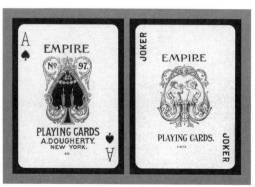

AD36

AD36a

AD36 EMPIRE #97
Andrew Dougherty, c1900. Another brand introduced by Dougherty around the turn of the century. The brand was in production for many years, but specimens are scarce. The earliest copies of the Empire brand had a Best Bower before the special Joker of AD36a was created.

AD36a EMPIRE #97
Andrew Dougherty, c1905. The later issue came with a completely redesigned Spade Ace and Joker.

AD37

AD39

AD39 HUNGARIAN #32
A. Dougherty, c1910. The Seasons pattern was a 32 card deck with German suit signs. This deck had the same fine quality stock, and the same back designs, as Climax. The 7 of Bells is the only card by which a collector can identify the maker if he is not fortunate enough to have the box.

AD37 CLIMAX #14
A. Dougherty, c1905. A fine quality deck printed on high priced paper, with a high gloss finish. This brand was also issued earlier with the AD15 Ace of Spades and a Jolly Joker.

AD38

AD38 DOUBLE PINOCHLE #78
A. Dougherty, c1905. The Pinochle versions of Climax were of the same fine quality. The double pinochle (64 cards) came in a tuck box with no mention of Climax. The front flap also said "finished with Pegulose". The 48 card deck came in a slip case marked 'Climax Pinochle'. Other than the sevens and eights the decks were identical. Minor variations of this Ace are known, for example, as pictured but without the words 'New York' in the banner.

AD40

AD40 MARGUERITE #130
A. Dougherty, c1910. A fine quality, Whist size card originally made with special backs, often of a floral design. Eventually, the brand was changed to make 'narrow named' cards to compete with the Congress #606W made by USPC. The cards were similar in most respects to this brand, which is not surprising as they were, by then, the same company. Narrow named cards are very collectible, especially by back collectors.

All of the known varieties of Marguerite Narrow Named have been listed. It should be noted that some of the titles have several variations and most come in more than one color combination. The Narrow Named cards came usually in flowers but scenic, people and animals were also made. The following backs are known and there are undoubtedly many more:

FLOWERS

Apple Blossom	Aster	Azalea	Black-eyed Susan
Carnation	Chrysanthemum	Clover	Daffodil
Dahlia	Gladiolus	Golden Iris	Goldenrod
Hollyhock	Hydrangea	Iris	Lily of the Valley
Lilac	Orchid	Pansy	Peony
Pink Rose	Poinsettia	Poppy	Purple Iris
Rose	Sunflower	Sweet Peas	Thistle
Tiger Lily	Tulip	Water Lily	White Lily
Windflower	Yellow Rose		

SCENIC

Afternoon	Cherries	Forest Stream	Full Sail
Heavy Seas	Lagoon	Mardi Gras	Sailing
The Canyon	The Harbour	The Lake	

PEOPLE OR ANIMALS

Anona	Black Beauty	Clown	Dancers
Debutante	Diana	Dolly	Dutch Boy
Flower Girl	Nanette	Nymph	Proposal Sprite
Surf Rider	The Huntress	The Pirate	The Scout
The Swan	Troubadour		

Private and special brands were an important part of the line of every major playing card producer. Many decks, with unusual names, have been found. Some will forever remain a mystery as to the actual maker, but there are many clues to help us in identification. These special brands, occasionally using fictitious company names, were often created for special purchases by large retail organizations. Special prices had to be offered to obtain this lucrative business and the manufacturers did not want their 'stock' brands being sold at below normal prices.

In attempting to identify makers of private brands there are a number of steps which might yield information. Firstly the tax stamp might identify the maker. Secondly, checking the court cards against the regular brands of the manufacturer can provide a strong indication because companies rarely made special plates for these orders. Thirdly, addresses and other data on boxes can give further clues.

The following brands have been conclusively identified as being made by Dougherty. It should be noted that the buyers of some of these brands might order them from more than one maker, for example, AD44 American Whist League, also had cards made by National and APCC of Kalamazoo. Radbridge, AD45, was also made by American Playing Card Co.

AD42

D42 MOON PINOCHLE #7
A. Dougherty, c1910. The pictured Ace of Spades was used for many special Dougherty decks, including Hockey #7 (the great Joker is pictured), Belmont #872, Rocket #276 and Pinochle #07.

AD43 SEBAGO #226 *(not pictured)*
A. Dougherty, c1910. This was a special brand Pinochle deck, identified by a dated tax stamp which came with the same Ace as AD43a.

AD43a

AD43a SEBAGO #225
A. Dougherty, c1910. A standard version was issued with 52 cards and an Indian Joker. The Sebago brand was one of the highest quality decks made by Dougherty.

AD41 MOON #1 *(not pictured)*
A. Dougherty, c1910. This special brand has an unusual Ace of Spades and Joker (a yellow moon).

AD41a

AD41a MOON #1
A. Dougherty, c1910. The same deck was made in a wide, poker size, version.

AD44

AD44 AMERICAN WHIST LEAGUE #109
A. Dougherty, c1895. Each year the League contracted with a manufacturer to supply a large number of decks. They were used in Whist boards for tournament play across the country. The Whist League also is known to have bought from National, Kalamazoo and American. As these decks were specifically for Whist, they were usually issued without Jokers. Interestingly, the deck pictured, which has the AWL back, also has the interesting Owl Joker playing with the AWL cards!

AD45

AD45 RADBRIDGE #201
A. Dougherty, c1910. A special brand made for and sold only by Radcliffe & Co., New York & London, a stationery chain. In some instances the Dougherty name was used on the Ace of Spades, while Radcliffe was used on others. Rad-Bridge was always on the Jokers.

AD48

AD48 'D' #8
A. Dougherty, 1909. A rare brand with a very unusual Joker, perhaps representing Dougherty's three sons! Note the prominence of the date of establishment. This brand was patented by W. H. Dougherty in 1909, two years after the sale to USPC in 1907, indicating that at least one of the sons stayed involved with the company for a period of time.

AD46

AD46 TURTLE
A. Dougherty, c1910. While the Dougherty name does not appear on the cards, the courts and design of the Ace of Spades strongly suggest this maker.

AD49

AD49 MONITOR
A. Dougherty, c1910. Another very rare brand of high quality which came in a gold embossed slipcase.

AD47

AD47 WALDORF #230
(formerly MSW134), A. Dougherty, c1915. The deck featured wide named, monotone backs. It is also known with 'No. 240' printed on the bottom of the Ace.

AD50

AD50 COMET
(formerly MSW74), A. Dougherty, c1905. This deck has since been proven to have been made by Dougherty.

The next three decks have very similar Aces and were undoubtedly manufactured by Dougherty as cheaper, generic brands to compete in specific markets.

AD51

AD51 PILOT #5
Andrew Dougherty, c1890. These cards are of Steamboat quality and with unmistakable Dougherty courts and a rather ornate Ace of Spades and Joker. The address on the box is 80 Centre St., New York. This deck also came without the brand name on the Ace in a no index version.

AD51a   AD51b

AD51a #000 STEAMBOAT
A. Dougherty, c1910. Clearly a later version of the Pilot deck. The only known copy has a 1916 tax stamp.

AD51b EAGLE CARD COMPY.
Andrew Dougherty, c1890. The same deck as AD51 with a different name on the Ace of Spades.

AD51c EAGLE CARD COMPY. (not pictured)
Andrew Dougherty, c1890. Eagle also made a square-cornered version with the AD51b Ace of Spades and Faro style courts.

AD52

AD52 WIRELESS #117
A. Dougherty, c1920. An interesting wide deck made with an anonymous Ace of Spades and a great Joker. Examination of the courts show it was made by Dougherty but the box has a USPC tax stamp from 1924-27. This deck was also issued in a narrow version with USPC courts.

AD53   AD54

AD53 PLATE ACE EUCHRE
c1865. The pictured deck was found in a cardboard box entitled "The Game of Bezique Complete" with three other 32 card decks. Printed on the inside bottom, underneath the lining, was "These cards are of the best American Manufacture and are retailed at One Dollar per pack". The cards were described as "A. Dougherty's best Double Head, Plate Ace Euchre packs made exclusively for A.B. Swift." Interestingly the courts appear identical to the Samuel Hart deck (NY26).

AD54 MANHATTAN WHIST CARDS
c1865. These cards, with later Netherland pattern courts, came in a wrapper printed "Manhattan Whist Cards – American Manufacture". The emblem on the wrapper includes an eagle, a crest similar to that of California and the word 'Excelsior'. While the evidence is inconclusive, we have opted to place this deck in the Dougherty chapter.

THE UNITED STATES PLAYING CARD COMPANY

In early January, 1867, A. O. Russell, Robert J. Morgan, James M. Armstrong and John F. Robinson Jr. formed a partnership and purchased from the Cincinnati Enquirer the job printing rooms which occupied the first and second stories at 20 College St., in Cincinnati. The firm commenced business under the name of Russell, Morgan & Co., Job Printers, under the active supervision of Messrs. Russell and Morgan. Not long after the new management took over, business had increased to such an extent that the firm had to seek new and larger quarters. They purchased a lot on Race St., and built a 20,000 square foot, four story brick building. The firm moved into its new quarters in November 1872. At the time, they believed it would be impossible to utilize all of their space, but were soon proven wrong!

Early in 1880, Mr. Russell proposed to his partners that they manufacture playing cards. At that time, eastern manufacturers like New York Consolidated Card Co. and Andrew Dougherty controlled the playing card business. Nonetheless, Russell believed "that this was a splendid field for operations on the part of men who would go into the business on liberal principles and only ask a reasonable price for their goods". The management agreed unanimously with the proposal and the project began. The first step was to add two stories to the building, making it six stories tall. Then, new and improved machinery was purchased and a very modern playing card plant was established on the two top floors.

From the beginning the company used only the most modern of equipment and they had many new machines designed and built expressly for their plant. One of these was a card-punching machine that enabled cards to be punched quickly and accurately from the printed sheets. This secured a greater degree of precision than the manual method of cutting and trimming cards used prior to that time.

The first pack of cards was produced on June 28th, 1881, and as Morgan handed it to Russell for inspection, he laughingly said "That pack of cards cost $35,000". They started on that memorable day with 20 employees, a strong determination to be successful and with the capacity to make 1,600 packs per day.

Two years later the company was again compelled to seek larger quarters. They built a new building, five times the size of the existing one and when it was completed, the company had the capacity to supply two-thirds of the demand for playing cards in America. As they continued to prosper and grow, they steadily increased their domestic market share, at the same time building a significant export business to England, Australia, Canada and many other countries.

Russell, Morgan & Co. also made labels, show cards, and over half of the circus posters for the United States, Mexico and Canada. By 1889, just 22 years after their start, and a mere eight years after producing their first deck, they had grown from a 20 employee, 1,600 deck a day firm, to one producing 30,000 decks a day and employing 630 people. And that was only the beginning!

When Russell, Morgan & Co. started in 1881, they had just five brands of playing cards. These were Tigers #101, Sportsman's #202, Army #303, Navy #303 and Congress #404. New brands were added in subsequent years to meet customer wants and to gain a competitive edge. Many variations of each brand were produced over the years as styles and customer demands changed. All the different Aces of Spades and Jokers that could be found have been separately identified and photographed with the exception of very minor differences caused by re-engraving of the plates.

Most, if not all, of their standard pattern backs came in either red or blue. In addition some brands had backs that also came in green or brown. These latter two colors are much scarcer in most brands and styles and are considered by many collectors to be more desirable. They thus command a higher price, all other features being equal. This is especially true, for example, with the Bicycle #808 brand.

The following information will help date playing cards made by Russell, Morgan & Co. and successor companies. First of all, in 1885 the Russell & Morgan Printing Company was formed to succeed Russell, Morgan & Co. Then, in 1891 the company name was changed to The United States Printing Co. following the acquisition of two other printing businesses. Finally, in 1894 the United States Playing Card Company was incorporated to hold all the playing card manufacturing operations of the, by now, greatly expanded activities of the United States Printing Co. which continued with all other aspects of the group's printing operations.

1881 to 1885 - the Aces of Spades read 'Russell, Morgan & Co.' or 'Russell & Morgan Co.' Probably, 'Russell, Morgan & Co.' came first. Hereafter they will both be abbreviated R&M.

1886 to 1891 - the Aces of Spades read 'Russell & Morgan Printing Co.' (RMP).

1891 to 1894 - the Aces of Spades read 'The United States Printing Co.' (USPn).

1894 to 1925 - the Aces of Spades read 'The United States Playing Card Co.' (USPC) or 'U.S. Playing Card Co., Russell & Morgan Factories' (USPC - RM Fact).

1926 to date – The United States Playing Card Co. (USPC).

US1a TIGERS #101
USPn, 1891. This issue carried the name of the brand and a tiger image on the Ace of Spades and came with a revised colored Joker.

It is not possible to precisely date any deck by its name alone because new Aces of Spades were not always designed when the company name was changed. For example, there is no known Tigers #101 Ace with the Russell & Morgan Printing name, even though subsequent Aces used later names and Tigers were available throughout that whole period. In addition, many decks made in the 1895-1925 period did not include 'Russell & Morgan Factories' on the Ace. It also should be noted that a deck with an early Ace of Spades might be found with a much later tax stamp. Decks sometimes remained in stock for lengthy periods, perhaps because new stock was placed in old cartons and shipped first. Unusual decks from, the 1870s, for example have been found with 1910 tax stamps, indicating perhaps a prototype deck removed from the factory at a much later date.

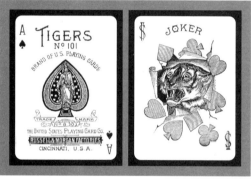

US1b TIGERS #101
USPC - RM Fact., c1894. A new Ace of Spades was made which was subsequently used as the basis for US1c and US1d. The pictured Joker was used for each of these three listings.

US1 TIGERS #101
R&M, 1881. Tigers #101 was their first brand. It was the least expensive deck presented as part of their opening line and had a generic R&M Ace of Spades (one without the brand name). A tiger head in a circle with the suits signs in smaller circles, printed in red and black, adorned the Joker. Tigers #101 were produced until about 1930 and five different Aces and three Jokers were used.

US1c TIGERS #101
USPC, c1905. It appears that in the later years, Tigers (as well as many other brands) came with both a USPC and USPC - RM Fact. (US1d) Ace of Spades.

US1d TIGERS #101
RM Fact., c1915.

One final note, almost all brands came with both plain and gold edges. The gold edged versions were usually identified by the addition of a small 'x' (extra) to the brand number. Exceptions, where known, will be noted.

The very first Tigers (and the first versions of several of the other early brands) were sold in attractive linen finish generic slipcases. These single colored slipcases, normally red, green, brown or blue, were embossed in gold or silver with the Russell & Morgan crest including the Statue of Freedom and the words 'U. S. Playing Cards'. After a year or two the company started to sell their decks in attractive tuck boxes or slipcases printed with the name and number of the deck and artwork appropriate to the brand. Up until the early 1900s the more expensive brands also had colorful wrappers sealing the decks within the boxes.

US2

US2 SPORTSMAN'S #202
R&M, 1881. The second highest grade of playing cards was introduced by Russell, Morgan & Co. in 1881. The first issue, Russell, Morgan & Co., had an interesting, picturesque Ace of Spades. This deck came with a retriever and rabbit-hunting scene Joker. A slightly later version, identical except it came with Russell & Morgan Co. printed on the Ace, used the US2a retriever and duck-hunting scene Joker. In addition, a sample book at USPC has a very different duck-hunting scene Joker. All subsequent decks had a retriever Joker, which was changed several times during the run of the brand. The Sportsman's brand usually came in a slipcase although the second deck referred to here was found in a fancy, colored tuck box.

US2a

US2a SPORTSMAN'S #202
RMP, c1886. The Ace of Spades was similar to US2, but far less detailed. The duck-hunting scene Joker came with this deck.

US2b

US2b SPORTSMAN'S #202
USPC - RM Fact., c1900. A new Ace of Spades was designed without any of the intricate detail of the preceding two. There are two versions, one with Russell & Morgan Factories printed across the bottom and one without. The pictured retriever Joker was used for this and subsequent Sportsman decks.

US2c

US2c SPORTSMAN'S #202 SERIES A
USPC, c1900. Another Ace of Spades for this brand, originally thought to be c1925, but now known to have been issued around the turn of the century. The 'Series A' referred to the size of the indices. The same Joker was used as in US2b. One mint version of this deck was found in its correct slipcase branded as Bridge Whist #88 with a 1901 tax stamp.

US2d

US2d SPORTSMAN'S #202
USPC - RM Fact., c1910. A newly discovered Ace of Spades with 'jumbo indices' similar to US34. It came with another version of the Retriever Joker.

US2e

US2e SPORTSMAN'S #202
USPC, c1920. This version has a generic Ace and Joker, but a Sportsman's box and back. Sportsman's was discontinued in 1936.

Armed Services Playing Cards were an important part of the Russell, Morgan & Co. line. Both Army cards and Navy cards were made under the style #303. They were made only for two years when split in demand caused the company to merge them as Army & Navy #303.

US3

US3 ARMY #303
R&M, 1881. The original Ace of Spades and Joker had battle scenes. Care should be taken when buying US3 or US4, as a fine reproduction of each was made for the 100th anniversary of the company in 1981. Interestingly, both Army #303 and Navy #303 came in a slipcase box with the printed Ace on the box upside-down.

US3a

US3a ARMY #303
R&M, 1883. The second Army #303 Ace of Spades was a simplified version of the original. It still had battle scenes, but without the intricacies of its predecessor. The colored Joker is a caricature of an early Army officer.

US3b ARMY #303 *(not pictured)*
USPC, 1981. Reproductions of the original Army and Navy decks were made in honor of the 100th anniversary of Russell, Morgan & Co. in 1981. They were produced in a limited edition of about 1,000 double decks, which were never placed on public sale. They were given as gifts to USPC's biggest customers including department stores and casinos. Close examination shows slight differences from the originals. The most telltale sign is the glossier paper. The reproductions each come in a slipcase with a rougher texture and the two slipcases come packed in a luxurious presentation package.

US4

US4 NAVY #303
R&M, 1881. The first Navy #303 Ace of Spades had battle details similar to its companion deck, Army #303. The Ace shows the battle between the Monitor and the Merrimac, with an early frigate in the background. The Joker shows the Monitor sinking.

US4a

US4a NAVY #303
R&M, 1883. Like US3a, the second Navy #303 Ace was greatly simplified. The Ace has two sailing ships and the Joker is a tipsy sailor.

US4b NAVY #303 *(not pictured)*
USPC, 1981. As stated in the description of US3b, an accurate reprint of this deck was made for the 100th anniversary of the company.

US5

US5 ARMY & NAVY #303
R&M, 1884. In 1884, in order to avoid any favoritism due to split demand between Army and Navy decks, as well as to reduce production and printing costs, Russell & Morgan Co. decided to combine the decks under the Army & Navy label. The company used the Navy style Ace of Spades for the first edition with the US3a Army Joker.

US5a

US5a ARMY & NAVY #303
RMP, 1885. With the start of Russell & Morgan Printing, a completely new Ace of Spades, Army & Navy #303, was created. The Ace had an eagle holding a '303' shield at the bottom of the card. The colorful Joker was redesigned, picturing an American eagle, sitting on an American flag that was draped on a globe, and holding an 'RM & Co.' banner in its beak.

While the RM Co. was producing US3, US4, and US5, they also made more luxurious versions of these brands. These featured gold edges and were numbered '505' instead of '303'. The #505 brand however, was eventually replaced by #303x (extra) which was noted on the box, but not on the cards. Some decks have also been found with #3032. It should be noted that special issues, made for large customers of USPC, had a '2' added to the style number. In some instances, these decks may have had a special Joker, box or back design.

Seven more Aces of Spades are known for the Army & Navy brand that was eventually phased out in 1931. All of these use another newly designed Joker picturing a battleship with a gun turret and an Army camp at the bottom. This Joker was modified several times throughout the run of the brand and used with different Aces of Spades (pictured with US5h).

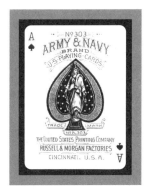

US5b   US5c

US5b ARMY & NAVY #303
RMP, 1885. At about the same time, another RMP Ace of Spades was designed. It retained navel vessels and had an eagle and flags at the bottom. It was the first to say 'US Playing Cards'.

US5c ARMY & NAVY #303
USPn, c1891. This deck utilized the same Ace as US5d, but for the Russell & Morgan Factories of the United States Printing Co.

US5d   US5e

US5d ARMY & NAVY #303
USPC – RM Fact., c1895

US5e ARMY & NAVY #303
USPC – RM Fact., c1900. The outline was dropped around the title and Russell & Morgan Factories moved to the bottom.

US5f   US5g

US5f ARMY & NAVY #303
USPC, c1905. A plain USPC Ace of Spades, similar to the previous listing, was used for this version.

US5g ARMY & NAVY #303
USPC, c1900. A simple USPC Ace which still retained the outlined 'Army & Navy'.

US5h

US5h ARMY & NAVY #3032
c1910. An example of a #3032 Ace made for Estanco de Naipes del Peru.

The first deck of Congress cards rolled off the presses of the Russell & Morgan Company in July 1881. It was the finest and most expensive of their first group of brands, and was a beautiful deck of playing cards, made of the finest enameled stock.

In the description of these prized cards, Russell & Morgan advertised in 1887:

No. 404 CONGRESS
For fine trade these cards have not their equal, being the "richest" goods on the market. Imagine a pack of cards with faces equal to the finest steel-plate printing, with a full portrait of "Dundreary" for the Joker, and for the Ace of Spades a perfect engraving of the Capitol at Washington. And then the backs! A rich blue, red or black for a background, throwing out a beautiful perspective in burnished gold, with which for elegance and tone nothing can compare.

Each Pack neatly wrapped, and in a green silk card case stamped in silver.

No.606 EXTRA CONGRESS
Extra Congress being the above, with pure Gold Edges, and in green silk cases stamped with gold, are really superb.

The first Congress decks featured lavish backs, done in a single color, with gold geometric designs and the Dundreary Joker. Their next version modified the Ace design, had the Dundreary Joker and introduced the stunning Japanese Lacquer Backs. Like the geometric designs they came in a single color with lavish use of gold or silver forming the design. All decks of this type and subsequent Congress decks came with gold edges and were described as #606.

Most of these Lacquer Back decks still incorporated the Dundreary Joker but an innovation that appeared with certain of these back designs was the introduction of a named Joker, i.e. the name of the back printed on a matching Joker. The Lacquer Back designs were never named on the back of the card, it is only when we get to the later, colorful, artistic designs that it is usual to find the name of the design written in small gold letters at the bottom of the card back. The exception to this practice is when the deck has a named Joker, thus eliminating the need to put the name on the card back.

The Lacquer back decks were issued concurrently, for a short period in the late 1890s, with the next innovation, the monotone backs. These were usually blue or beige with a scenic view. The final and most common type of wide Congress backs was the pictorial back that was produced for a period of approximately 25 years starting in 1899. These came in a myriad of designs - women and children, scenes and buildings, animals and flowers, initials and patterns, famous people and paintings – just about everything was, at one time, featured on a deck of Congress cards.

How highly the company valued these quality cards is indicated by their 1888 wholesale price list where their cheapest brand, Steamboats, were $.50 a dozen or $6.00 a gross and the Extra Congress were $7.00 a dozen or $84.00 a gross!

With the increasing popularity of Bridge, USPC decreased their wide Congress output and by the 1927 they had phased these decks out. They had already introduced the very popular narrow variety with many wonderful backs that have continued until the present.

US6

US6 CONGRESS #404 and #606
R&M, 1881. Congress cards were originally issued in two variations - the #404 brand with plain edges which was dropped in the 1890s; and the #606 (gold edges) style remained as the top of the line. We have pictured two slightly different 'first' Aces - 'Russell, Morgan & Co.' and 'Russell & Morgan Co.' In addition we show three versions of the Dundreary Joker, the first without 'US' in the corners, the second with 'US' added and the extremely rare bordered style with flesh tones.

US6a

US6a CONGRESS #606
RMP, c1885. The RMP Ace of Spades, with the Capitol Building illustrated in far less detail than on US6. The Dundreary Joker was retained through the USPn era.

US6cc

US6cc EXTRA CONGRESS #606
USPC – RM Fact., c1895. A very high quality, gold edged edition that is quite rare.

US6b

US6b CONGRESS #606
USPn., c1891. The US Printing Co. Ace of Spades also mentioned the Russell & Morgan Factories.

US6d CONGRESS #606 (not pictured)
USPC, 1899. Around this time USPC made some Congress decks with an early USPC Ace of Spades with no mention of RM Fact., yet dated 1899. This type of anomaly is bound to cause a problem when trying to positively date a deck that is found without its box. The Ace is not pictured, as it is the same as US6c, without the RM Fact. reference.

US6e   US6f   US6g

US6e, US6f and US6g CONGRESS #606
USPC – RM Fact., 1899. These three Aces of Spades were used from 1899 to 1922. US6e was usually dated. All three of these decks are quite common, although there are certain backs in great demand that command high prices.

US6c

US6h

US6h CONGRESS #606W
USPC - RM Fact., 1917. An early USPC narrow deck named Congress 606W (Whist Size). The whist size Congress cards were introduced in 1917 and were made concurrently with their wide counterparts until 1927 when the wide were discontinued. These cards also had scenic backs. The 'W' on the Ace was dropped after a little while as the wide issues decreased in popularity.

US6c CONGRESS #606
USPC – RM Fact., c1895. The earliest USPC, Russell & Morgan Factories Ace of Spades. About this time USPC started using either a named Joker or the common 'Capitol' Joker. Two named and the Capitol Joker are pictured.

US6i   US6m

US7a   US7a-j

US6i CONGRESS #606W
USPC, c1921. This Ace dropped the reference to Russell & Morgan Factories.

US6m CONGRESS #606
USPC, 1930 on. The Ace of Spades went through many minor variations to arrive at the one used today. These have not been shown, but we have pictured the Joker of a late 1930s Congress deck.

US6n #666 MULTI PLAYING CARDS (not pictured)
USPC - RM Fact., c1904. A very unusual and rare Congress deck consisting of 53 different Congress backs. These decks, which were produced for only a relatively short time, were designed to provide potential customers with samples of Congress quality cards and for use in playing Solitaire. The Ace of Spades that comes with this deck is US6g.

Steamboat #999 was the first new brand to be added to the original Russell & Morgan line. It was introduced in 1883, and was the cheapest deck they offered for sale. Steamboats were added to meet the competition as all other playing card manufacturers had a 'Steamboat' brand, usually one of the cheapest, retailing for as little as 5 cents per pack.

US7a STEAMBOAT #999
USPn., 1891. This Ace identified the brand with the '999' and an engraved steamboat, but still did not mention the Steamboat name.

US7aj STEAMBOAT #999
USPn., 1891. Again, a special watermelon Joker sometimes (always?) used with this Ace.

US7b

US7b STEAMBOAT #999
USPC - RM Fact., c1895. The first Russell & Morgan Factories Ace of Spades was designed in 1895.

US7   US7-j

US7c   US7d

US7 STEAMBOAT #999
R&M, 1883. The Ace of Spades, similar to 'Tigers' (US1), had no brand name and used a generic R&M Ace. In the original Encyclopedia it was stated that the Joker for all Steamboat decks issued by Russell & Morgan and USPC, except for US7-j and US7aj, was a steamboat with smoke emitting from the two large stacks near the front of the ship. While this may be accurate, we have never seen a US7 or US7a deck with the steamboat Joker. They have all included one of the two special jokers listed as US7-j and US7a-j.

US7-j STEAMBOAT #999
R&M, 1883. The special 'watermelon' Joker issued with some (all?) of the US7 decks.

US7c STEAMBOAT #999
USPC - RM Fact., c1910. A later RM Fact. Ace of Spades.

US7d STEAMBOAT #999
USPC, c1900. The first USPC Ace, a simplified version of US7b, probably available at the same time, as USPC seemed to have both a RM Fact. version and a plainer USPC version of many brands available simultaneously.

US7e STEAMBOAT #999 (not pictured)
USPC, c1925. A revision of US7c dropping reference to RM Fact. and still in use today.

AD CARD

BICYCLE BACKS

In 1885, Russell & Morgan Printing Co. introduced their famous Bicycle playing cards. They were then advertised as 'The cheapest enameled Playing Cards in the World'. Bicycle cards were not only a huge success, but became the best known brand in the world. They started with only one back, Old Fan, in either red or blue, but eventually finished with a total of 82 different back designs. Many backs were short-lived, Old Fan for example was dropped in 1894, but some continued longer and the well known Rider back is still in use after 112 years. While not nearly as scenic or historic as the

Congress backs, the variety of Bicycle designs has made them a favorite of collectors. A very fine collector's handbook was issued by Mrs. Joe Robinson in 1955 which pictured all the backs. They were also pictured in Clear the Decks, Vol. VI, No.2.

Some examples of early bicycle backs, which had bicycle, motorcycle and automobile motifs, are shown above. Also pictured is an advertising card that was distributed to promote the Bicycle brand.

US8

US8 BICYCLE #808
RMP, c1889. While originally thought to be the first Ace of Spades for this brand, it is now known to be the second. It originally came with a high-wheel Joker with 'US' in the corners, but subsequently the two low-wheeler Jokers pictured were issued with this Ace. The one with the King riding is extremely rare and is printed in blue, yellow and red. The other, with the contemporary gentleman riding a new style low wheel bike, was only issued for a few years and is scarce in its own right. In addition to the three standard Jokers used with this Ace, we have pictured a special advertising Joker which came with a gold edged Lotus back standard deck in a leather and celluloid case also advertising Krupp No.1. This is the only instance we can find of USPC allowing their Bicycle brand to be used to promote another company's product.

US8a

US8a BICYCLE #808
RMP, 1885. This version is now known to be the first Bicycle Ace of Spades. Neither of the first two Bicycle Aces used the number '808', although the number was used in their promotional materials. The Joker was a high-wheeler with Best Bower written at the top.

US8c

US8c BICYCLE #808
USPC - RM Fact., c1905. A later, and certainly the most common of the early Bicycle Aces. This Ace was used until at least 1925. The Joker is similar to US8b but the King wears a crown. This Ace, it should be noted, was used for the four, rare 'War Series' backs (two shown) produced in 1918 for a very short period and featuring instruments of war.

US8aa

US8aa BICYCLE #808
USPn., c1891. This Ace, discovered since the original Encyclopedia was issued, answers the question of whether United States Printing Co. published Bicycle decks with a USPn. Ace of Spades. The deck came with the 'low-wheel' Joker.

US8d

US8d BICYCLE #808
USPC, c1900. A plain USPC version of (US8b) with no mention of Russell & Morgan Factories was issued around 1900. As many of the backs were issued with several different Aces, in the absence of the tax stamp, the Ace is the only clue to dating a deck. The deck was printed with the Jester Hat Joker of US8b. This Ace was not used for a long period, perhaps only a few years, as we have seen many Bicycle decks issued from 1910 to 1925 but always with the US8c Ace.

US8b

US8b BICYCLE #808
USPC - RM Fact., c1895. The earliest Russell & Morgan Factories Ace of Spades. The Joker depicted a 'King', with a jester type hat, riding a bicycle past an 808 milestone.

US8e

US8e BICYCLE #808
USPC, c1930. The last of the standard Bicycle Aces is still in use today, albeit with minor variations.

US8f

US8f BICYCLE BRIDGE #86
USPC - RM Fact., c1920, French (Whist) Size. This was the first of a series of narrow decks in which the company cashed in on the already popular Bicycle name. Note that the numeral '86' is now shown on the milestone on the Joker.

US8fa BICYCLE BRIDGE #86 *(not pictured)*
USPC - RM Fact., c1922. This deck differs from the previous one by omitting the words 'French (Whist) Size'.

US8g BICYCLE BRIDGE #86 *(not pictured)*
USPC, c1925. A later version of this narrow deck was introduced when the RM Factories name was dropped.

US8h

US8h BICYCLE #88
USPC, c1930. The Bicycle version of the Jumbo index style (refer US34).

US8i BICYCLE PINOCHLE #48 *(not pictured)*
USPC, c1925. This deck was issued to replace the earlier version of Bicycle Pinochle #48 issued by USPC - RM Factories and listed below as US8ia.

US8ia   US8j

US8ia BICYCLE PINOCHLE #48
USPC - RM Fact., c1920. The earlier version of US8i was issued to use the Bicycle name for this popular game.

US8j BICYCLE JUMBO #808
USPC - RM Fact., c1920. A novelty deck twice as long and twice as wide as a normal whist sized deck.

US8k

US8k BICYCLE BRIDGE 888
USPC, c1925. The Bicycle Bridge multicolor series was issued around 1925 to capitalize on the Bridge craze. The decks came with the standard Joker, beautifully colored, and tinted edges in a fancy box with art deco backs. The Ace pictured is a Canadian version.

The comment in the previous listing about Canadian versions needs some discussion. USPC opened a plant in Toronto, Canada in 1914 and moved to Windsor, near Detroit, in 1918. They made a number of the popular brands in the Canadian plants which remained in production until the early 1990s. A more complete description of USPC's Canadian operations and descriptions of their cards are included as part of Chapter 16, Canadian Standard Playing Cards.

US9

US9 TOURISTS #155
RMP, 1886. Tourists #155 was one of the new brands added by the Russell & Morgan Printing Co. At this early stage of their existence, the company's rapid growth made it necessary to compete at every price range with the other manufacturers. Tourists were graded between Tigers and Sportsman's. Quoting from an 1887 advertisement, "They are a happy medium between unenameled cards (Tigers and Steamboats) and enameled cards (from Bicycles upward), having the finish of the former and the same style backs as the latter. Each pack in a neat tuck box."

US9a

US9a TOURISTS #155
USPC - RM Fact., c1895. This Ace of Spades was accompanied by a similar, but now black and white, Joker.

US9b

US9c

US9b TOURISTS #155
USPC - RM Fact., c1910. The same Joker was used as in US9a. The brand was phased out in 1931.

US9c TOURISTS #155
USPC, c1900. This plain USPC version appears to have been manufactured, like the early USPC Aces of several of the other brands, around the turn of the century.

US10b CAPITOL #188 *(not pictured)*
USPC - RM Fact., c1910. The Ace is similar to, and later than the following two RM Factories versions. It has the name in plainer print with 'No. 188' below the name.

US10c

US10c CAPITOL #188
USPC - RM Fact., c1898. The first Russell & Morgan Factories Ace for this brand was issued in the late 1890s.

US10

US10 CAPITOL #188
RMP, c1886. This was another of the new brands introduced during the Russell & Morgan Printing Co. era. This brand fit between Bicycle and Sportsman's. To quote an early advertisement, "Capitols are superior to Bicycles because they are composed of better stock, being stronger and more durable, and will, of course, last considerably longer. In this grade we offer six beautiful backs, in pink, blue, and buff enamel". The pictured Capitol Joker was used throughout the run of the brand.

US10d CAPITOL #188 *(not pictured)*
USPC - RM Fact., c1905. A slightly different version (not pictured) of US10c with the '188' moved below the name. The brand was phased out in 1928.

US11

US11 SQUARED FARO #366
R&M, 1887. This deck was made to compete with the eastern manufacturers supplying the gambling casinos. The cards had squared corners and were one of the two known USPC decks made with no indices (see US41). To quote from an introductory advertisement, "Faro dealers know when they get a good pack of dealing cards, and all who have tried ours, pronounce them superior to all others. They are square (in more senses than one). They will fit any dealing box. There is margin enough on the faces to trim many times. They are hermetically sealed to prevent their being tampered with. There are three sizes: Eights, Nines, and Tens" (referring to the exact thickness of the cards). Although the brand was first produced during the RMP era, it was still sold until 1900. The Russell & Morgan Co. Ace of Spades was used during the entire life of the brand.

US10a

US10a CAPITOL #188
USPC, c1895. Basically the same Ace as US10, but modified to read USPC. Although early, it does not mention RM Fact. and the #188 was a new feature. The Joker was modified slightly by placing the 'US' initials in the two corners.

US12

US12c

US12 CABINET #707

RMP, 1888. Introduced early in 1888, Cabinet #707 was originally made only as a 32 card deck, intended for Euchre. An early advertisement read, "Cabinet Playing Cards are made especially for progressive euchre, and each pack contains 32 cards, counters, and a description of how to play the game. They are highly enameled and beautifully finished, having six handsome backs, so that in playing progressive euchre, no two tables need have the same backs. They are contained in fine cloth cases, and can be had either plain or with gold edges (#707x)." The four markers, two red threes and two black sixes, were cards with black backs (pictured) advertising the company. Each deck was packaged in a colorful wrapper inside an embossed box with the markers and an instruction leaflet. The deck came with the pictured Joker although it may have been included only in the later decks.

US12c CABINET #707 U.S. WHIST SIZE

USPC - RM Fact., 1906. The Cabinet brand now changed to a narrow width for use in Whist and Bridge and was continued until about 1930.

US12d

US12d CABINET #707x

USPC, c1899. A wide, gold edged version which came in an attractive wrapper within a slip case and used a plainer USPC Ace.

US12a

US12a CABINET #707

RMP, c1890. With the waning popularity of Euchre, Cabinet, which had found its niche in the Company's line, was changed to a regular 52 card deck. It was slotted between Capitol and Sportsman's, and featured monotone, scenic backs in a variety of colors. The Joker was completely redesigned and it was this Joker which was used for the remainder of the life of the brand as a wide issue. Cabinet was discontinued in 1930.

US12b

US12b CABINET #707

USPC, c1898. This deck, an early issue, did not have RM Fact. mentioned on the Ace of Spades.

US12e

US12e CABINET BRIDGE #707

USPC, c1925. A narrow Jumbo index deck almost identical to US35 Auction Bridge #708.

US12f AUCTION BRIDGE #707 *(not pictured)*

USPC - RM Fact., c1910. This 'U.S. Whist size' deck (same as US12c) came, interestingly, in a #708 Auction Bridge slipcase. As the deck was unopened we know it left the factory this way.

US13

US13 TEXAN #45

RMP, 1889. In July 1889, RMP added another new brand, Texan #45. To quote the initial advertisement, "There are card players who prefer to use a card that is not enameled, but have been compelled to used enameled cards for the simple reason that cards like Tigers and Steamboats were not obtainable with a high enough finish and strength to answer the purpose. We have therefore supplied this deficiency in the Texan '45. These cards are of superior strength and quality of stock, evenly and highly finished, and are the best wearing unenameled cards made". They were graded between Bicycles and Capitols.

US13-1

US13a

US13-1 TEXAN #45

USPn., c1891. A redesigned Ace was introduced when the company name changed in 1891. Unlike many US Printing Aces it also mentioned Russell & Morgan Factories. The Joker for this and subsequent versions was essentially the same as the original with the addition of 'US' in the corners.

US13a TEXAN #45

USPn., c1893. A slightly later USPn. - RM Factories Ace was introduced before the change to USPC in 1894.

US13b

US13b TEXAN #45

USPC, c1900. The USPC version was likely issued around the turn of the century. The brand was phased out in the United States in the 1920s but continued in USPC's Canadian division until recently.

US13c TEXAN #45 *(not pictured)*

1984. In 1984, Philip Morris produced a set that included two decks of Texan #45 cards and a booklet on Poker packaged in an imitation leather slipcase. The Ace and Joker bore no resemblance to the original cards but the R&M Printing Co. name was used on the Ace and the cards themselves are excellent reproductions of the original cards. The lone star back was used for these decks.

The last new brands to be introduced in the Russell & Morgan Printing Co. era were presented with the following advertisement, "At the request of numerous German friends who attended the Skat Congress, held in Milwaukee, Wisconsin in June 1888 (where we were represented), we have prepared and placed in the hands of the dealers what we call American Skat cards, thereby filling a want on the part of those Germans who have emigrated to this country to cast their lot with us, as Americans. We have two types of these cards, one with German faces (after Wust's designs), and one with American faces." Both of these types were discontinued in 1901.

US14

US14 SKAT #2 GERMAN FACES

RMP, 1889. A 36 card deck (some with only 24) manufactured in two grades, one unenameled (SKAT #1) and this deck (SKAT #2) which is double enameled. The designs featured the German Eagle and the American and German flags entwined. The unenameled brand was dropped after a very short time. The same Ace was used for issues in the 1890s with the USPC name

US15

US15 SKAT #4 AMERICAN FACES

RMP, 1889. These decks had 52 cards plus a Joker. Our copy is in a box labeled '#4x Special Whist Playing Cards'. Skat #3 was the unenameled number. The full pack was to allow Skat players to play other American games. Neither American Skat #3 or #4 appeared in price lists by the turn of the century.

In 1891, Russell & Morgan Printing Co. changed its name to the United States Printing Co. As can be seen, some Aces of Spades were quickly changed to promote the new name, while some were retained as is. The Company grew rapidly and continued to introduce new brands including the popular Ivory series. These decks were made longer and narrow than standard cards.

US16

US16 IVORY #93 SERIES
USPn. - RM Fact., c1893. Under the U.S. Printing Co., the new "Top of the Line" extra whist size cards were introduced. They were double enameled, and made of fine linen stock.

US16a   US16b

US16a IVORY WHIST #93
USPC - RM Fact., c1898. The growing popularity of narrower cards for use in the game of Whist made this brand extremely successful.

US16b IVORY WHIST #93
USPC - RM Fact., c1905. The many changes in the Ace of Spades in this series attest to the popularity of the brand. The Elephant (Ivory) Joker was retained throughout the run of the brand.

US16c

US16c IVORY BRIDGE #93
USPC - RM Fact., c1906. This was the natural successor to Ivory Whist. The brand was popular amongst Whist players who were gradually converting to Auction Bridge. The Ivory brand retained its popularity throughout the Auction era, but disappeared with the coming of Contract Bridge in the late1920s. Of course, by this time Congress Whist size cards had established themselves as one of the leading narrow brands.

US16d   US16e

US16d IVORY PINOCHLE #930
USPC - RM Fact., c1895. The special size of the Ivory brand cards was adaptable for games requiring the holding of large number of cards in one's hand, which made them a natural for the game of Pinochle.

US16e IVORY PINOCHLE #930
USPC - RM Fact., c1900. Around 1900 USPC introduced a more modern version of the Ivory Pinochle deck. It comes with either 'No. 930' or '930' on the top of the Ace.

US16f   US16g

US16f IVORY PINOCHLE #930
USPC - RM Fact., c1910. This brand was popular for many years as the third RM Factories Ace attests.

US16g IVORY PINOCHLE #930
c1923. A later USPC version of this brand was produced when the company stopped using the RM Factories name in the early 1920s.

US16h BRIDGE #93
USPC - RM Fact., c1915. The successor to US16c continued to use the Ivory Joker.

US16h

Near the end of the USPn era, USPC made their first major acquisition. Arrangements were made in 1894 to purchase the National Playing Card Co., of Indianapolis. National, at that time, was producing about 20 brands of playing cards. As can be seen in the National section, some were discontinued, some were retained and some were adopted and continued under the USPC name. At the Columbian Exposition in Chicago, in 1893, the US Printing Company had a booth together with National. Both companies won medals for their excellence in workmanship.

It was around this time that USPn also acquired the business of Standard Playing Card Co. of Chicago. Then, to add to their already dominant position, in July 1894, they agreed, during a meeting to discuss the joint purchase with Dougherty and NYCC of Perfection Playing Card Company, to buy NYCC from the descendants of the Cohen family. After negotiations, NYCC agreed to the acquisition with the proviso that they continue to operate as a separate entity.

So in 1894 the United States Playing Card Co. was incorporated and the playing card operations transferred to it. The US Printing Co. would continue to make posters, labels, etc. while USPC, with Perfection, National, NYCC and Standard in the fold, would concentrate on the manufacture of playing cards.

US18 TREASURY #89 (not pictured)
USPC. - RM Fact., 1895. The original Encyclopedia stated that Treasury #89 was first issued in 1895 using a USPC - RM Fact. Ace of Spades and this deck is listed here as US18. However close examination of the picture revealed it was made by USPC and it is now pictured as US18-2 below. We cannot state definitely that US18 as originally listed actually exists.

We do know now that Russell & Morgan Printing Co. introduced Treasury #89 in 1890 as a new "Top of the Line" brand. It was advertised as "double-enameled and made of the finest linen stock" and designed especially to appeal to Poker players and Poker clubs. Two versions of this deck produced before 1895 are now known and are listed below as US18-1 and US18-2. The Joker, and indeed the Joker for most versions of this brand, used the one pictured under US18-2 depicting the Treasury building in Washington, an eagle and a cornucopia of coins and treasury notes.

US18-1

US18-1 TREASURY #89
RMP, 1890. A special Joker depicting coinage of that year accompanied the first Treasury Ace. This version is extremely scarce and only a couple of copies are known.

US17

US17 AMERICANISCHE GAIGEL
USPC, c1895. This was a follow up to the Skat cards with German faces (US14). It catered to the players of the second most popular game among German-Americans, Gaigel, which was the German name for Pinochle. The deck consisted of 48 cards, 2 each of the 9 through Ace. The cards were the same as those in the Skat deck except that the Ace of Spades (leaves) said Gaigel in place of Skat.

US18-2

US18-2 TREASURY #89
USPn. - RM Fact., c1892. A US Printing version of Treasury was produced a few years later with the newly designed Joker.

US18a TREASURY #89 (not pictured)
USPC - RM Fact., c1898. The Treasury brand continued its success and a short time later the Ace of Spades and Joker were redesigned. The brand was divided into two series - the regular series and the Club Series, both of which had floral backs printed in multi-colors as well as more conventional backs. Certain aces of this listing were issued without mention of RM Factories.

US18b

US18b TREASURY #89 CLUB SERIES
USPC - RM Fact., c1898. The Club Series Ace was almost identical to US18a.

US18c TREASURY #89 SERIES A (not pictured)
USPC - RM Fact., c1899. This revised edition of Treasury cards, Club Series, was designed principally for Poker playing and had extra large indices.

US18d TREASURY #89 SERIES B (not pictured)
USPC - RM Fact., c1899. This deck was identical to US18c in every way except it had four of the extra large poker indices. They were advertised as "for left and right handed Poker players".

US18e   US18f

US18e TREASURY #89
USPC - RM Fact., c1910. This was probably the last Ace of Spades for the Treasury brand. For an unknown reason, its popularity diminished and the brand was phased out before 1920.

US18f TREASURY #89 CLUB SOLO
USPC - RM Fact., c1900. Another Ace of Spades almost identical to US18c and US18d but this time headed Club Solo. Similar decks were produced for Euchre, Skat, etc.

US18g

US18g TREASURY #89 CLUB SERIES
USPC - RM Fact., c1900. Another version of the Club Series Ace has turned up. Our copy is in a box marked 'second quality' and has a special Ace of Clubs marked 'Treasury Seconds'.

US18h

US18h TREASURY #89
USPC, c1905. This Ace made no mention of RM Factories.

US18i

US18i TREASURY #892
USPC - RM Fact., c1905. Another version, this time noted as Treasury #892, was issued about the same time. The addition of the '2' to the brand number probably referred to a special order. Our deck comes, for example, in a slipcase advertising the United States Printing Co., makers of labels and folding boxes.

In 1895, USPC decided to create several exciting new brands, all of which would feature unusual court cards. There were several Stage decks (SE2, SE3, SE4 and SE5). They also produced, along similar lines, three transformation decks, Vanity Fair (T11), Hustling Joe (T9 & T10), and Ye Witches (T12). Finally, they produced the new designs described below.

US19   *US19a*   *US19b*

US19 TROPHY WHIST #39
USPC, 1895. The first of the decks with 'modernized' courts was produced in a Whist size deck. It came with an attractive trophy Joker. It is interesting to note that this series of new style decks, produced shortly after the incorporation of USPC, did not mention Russell & Morgan on their Aces. Besides being distributed in a Trophy Whist slipcase, this deck was available in a tuck box marked National Method Duplicate Whist.

US19a TROPHY WHIST #39
USPC, c1896. As has always been the case with new features on playing cards, many card players are reluctant to embrace new concepts. This explains the printing and simultaneous distribution of the same brand (with minor variations in the Ace) with standard courts.

US19b TROPHY BRIDGE #39
USPC, c1905. A later edition of US19a designed to follow the trend from whist to bridge. The Ace is the same as US19a except that 'Bridge' replaces 'Whist'.

US20a

US20 NEW ERA #46 *(not pictured)*
USPC, 1896. The issue made for the new era of USPC. What better way to announce the growing importance of this new playing card company than to create an extremely beautiful deck. The advertisements called the deck "colorful and created out of European Fashions".

US20a NEW ERA #46
USPC, 1896. A slightly different version of the New Era deck, with much plainer writing on the Ace of Spades and darker courts was issued, perhaps a few years later. This brand and Circus #47 were advertised for many years as high-end brands in the USPC price lists. New Era was still listed in a 1908 catalogue and Circus as late as 1925. Their scarcity and rarity today indicate that these novelties were never big sellers.

US21

US21 CIRCUS #47
USPC, 1896. Another deck produced at the same time and with a similar purpose to US20. To quote an advertisement, "The staid old Kings, Queens and Jacks have given way to various well-known ringmasters, clowns, and queens; dashing Circus designs."

US21a CIRCUS #47 *(not pictured)*
1896. The USPC was quick to acknowledge errors. The 'P' in Playing Cards ran into the Spade of the corner index on US21 and it was corrected in this edition.

US22

US22 MYSTIC #888

USPC, 1898. This most enterprising company attempted to corner the business of the large fraternal orders in the U.S. by creating a new brand catering to their needs. Their ad, "A fine line of Secret Society Cards. Special emblematic backs and Jokers: Back designs - Goat (General), Knights of Pythias, Elks, Shriners and Odd-Fellows. Fine linen stock, highly enameled and finished; gold stamped telescope cases. Put up solid or assorted, six packs to a carton". The scarcity of existing examples of these cards seems to indicate that the idea was not an overwhelming success. Examples of two Aces (they are all identical except for the name), three Jokers and the five backs are pictured.

US23

US23 PINOCHLE #48

USPC - RM Fact., c1898. These cards were advertised as 'Enameled, Bicycle quality'. The brand number '48' was an appropriate number, indicating 48 cards in the deck.

US23a PINOCHLE #48

USPC - RM Fact., c1910. A later version was produced to replace the more elaborate Ace of US23.

US24 PINOCHLE #64

(not pictured) USPC - RM Fact., c1898. The same brand was produced including the 7's and 8's. The #64 on the Ace of Spades indicated 64 cards in the deck.

US23a

The United States Playing Card Co. price lists, by this time, included the brands of the National Playing Card Co. Those brands, which were continued under the National name, are included in the National chapter. Reproduced here is part of a letterhead of the United States Printing Co., showing five of its plants. The Russell & Morgan #1, top left, and National #5, bottom left, produced most of the playing cards.

The next major move of the U.S. Playing Card Co. was to new quarters in Norwood, Ohio in 1901. Norwood, a suburb of Cincinnati, is where the company is still located today.

US27

US27 PETITE #21
USPC - RM Fact., c1909. The Ace of Spades in this brand was issued both with and without the company name although there was no obvious effort to hide the manufacturer in the latter case. The deck was the same size and quality as Junior #21, but the Ace of Spades and Joker were very different. Perhaps the company wanted a brand name more descriptive of the type of card?

US25

US25 NORWOOD #85
USPC - RM Fact., c1909. This deck, in modified Whist size, is thought by many who have seen it to be the most beautiful ever made by USPC. It features a colorful Ace of Spades, decorated aces, unique courts and two beautiful back designs. It does not appear to have ever been listed in a catalogue, or to have been advertised for sale. It might have been made as a presentation deck when they moved to their new location in 1901, or perhaps it was simply a later experiment that proved to be too expensive to manufacture in quantity due to the 12 color printing. The only records of this issue in the USPC files are in 1909 files and indicate that the entire issue was destroyed and never placed on sale and until more are discovered we will leave the date of manufacture as 1909. Only a few copies are known to remain and, based on their tax stamps, all appear to have left the factory in 1919! The known copies come in either a Bijou #1 box or a pink Bijou style box and the two backs, Cupid and Psyche (The Awakening) and Paul and Virginia (The Storm), were also used in National PCC Bijou decks.

US28

US28 INITIAL #54
USPC - RM Fact., 1909. In 1905, one of the decks introduced by the Congress brand was an 'initial' back. This proved to be most popular, and retailers had many requests for specific initials. USPC therefore decided to give the deck its own brand name and number. Interestingly, the wide Initial 54 seems to have been phased out by 1920 at which time all initials could be ordered on Congress 606W backs, framed in a stylish art nouveau design.

US26

US26 JUNIOR #21
USPC - RM Fact., 1892. This slightly smaller brand (2 1/2" x 3 1/4") was copyright in 1898. It was made in response to the demand for a size that would facilitate the holding of many cards comfortably. The same stock was used as in the Bicycle brand and they were often referred to as 'Bicycle Juniors'. Note the bicycle on the Ace of Spades and Joker. Certain of the backs used in the Bicycle brand (e.g. Lotus, Rider and Safety), were used for Juniors. Despite being offered for many years, this brand is very difficult to find. In the later years (we have seen it advertised in the 1920s) it was redesigned with a simpler Ace.

US28a

US28a INITIAL #54
USPC - RM Fact., c1912. A second, newly designed Ace of Spades always meant a good selling brand. The same Joker was used with both Aces.

US29

US29 CANTEEN #515

USPC, 1898. This brand was created for the Spanish-American War (see W47). It was a non-enameled, inexpensive deck, which was made to be sold to the servicemen at the Armed Forces canteens. After the war, the brand was dropped and we have only seen one copy, unfortunately without a Joker.

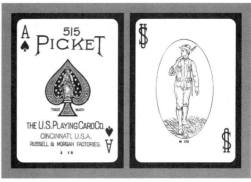

US30a

US30a PICKET #515

USPC - RM Fact., c1917. Obviously the brand was successful enough to necessitate a re-engraving. Perhaps, this was due to the fact that these decks were made available at a very low price to our fighting men. The Joker was changed although the motif remained the same. Interestingly, these decks were also available in a high quality, gold-edged version.

In 1907, after more than 65 years as an independent maker, the Andrew Dougherty company, operated by Dougherty's three sons, decided to sell out to the United States Playing Card Co. They also agreed that they would not re-enter the playing card business for 15 years. USPC operated the company as a separate entity until June 4th, 1930, when it was merged with NYCC to form the Consolidated-Dougherty Company.

Dougherty continued to manufacture new brands after the USPC purchase of its shares. Occasionally, USPC courts were used in a Dougherty deck, and many of the progressive ideas of both firms were interchanged. We also know that USPC decks were made at the Dougherty plant and vice versa. Many of the listings that follow are for new brands designed subsequent to the Dougherty purchase.

US31

US31 VICTORS #79

USPC, 1900. This brand was issued in honor of the victory in the Spanish American war (see W46). While it was previously believed to have been issued after WWI, we have a copy with a 1902 tax stamp. Note the addition of London to the Ace of Spades on the pictured Victors Ace. Readers may remember that Victor Mauger had returned to the United Kingdom in the early 189s as the representative of USPC and had built up a substantial trade in their playing cards through the American Playing Card Agency. After his death, around the turn of the century, USPC opened its own sales office in London.

US30

US30 PICKET #515

USPC - RM Fact., 1914. Another deck for soldiers was introduced in 1914 using the same style number as Canteen (see W48). The name was changed to Picket and the Joker portrayed a 'Doughboy' of World War I. As in the case of Canteen, the brand was dropped shortly after Armistice.

US31a

US31a VICTOR #79

USPC, 1900. The same brand with a different Ace of Spades, and, the 's' removed from the name on the Ace and Joker.

US32

US34

US32 PENNANT #252

USPC - RM Fact., c1910. Pennant was issued as a 48 card Pinochle deck, produced and marketed c1910. A 52 card version, complete with Joker has also been found. Other than this, little is known of the brand.

US32a PENNANT #253 *(not pictured)*

USPC - RM Fact., c1912. A narrow version of US32 came with a narrow version of the Joker. This appears to be one of the earlier USPC narrow brands that became so popular in the late 1910's and 1920's.

US34 JUMBO INDEX #88

USPC - RM Fact., 1895. Cards with Jumbo indices were introduced in 1895 and are still going strong today. The Ace of Spades had a great many titles, but the number '88' was constant (it refers to the extra large indices). All decks bearing the number 88 are included under this reference number. Other decks with Jumbo indices, but bearing a brand name and number are listed under the appropriate reference number (e.g. US2d and US12e). Over the years a variety of Jokers were used, including ones from other brands. For example, the pictured #88 Bridge Ace of Spades came with the Sportsman's Joker (US2d).

US33

US35   *US36*

US33 HELMET #119

USPC RM Fact., c1908. Helmet is another little known brand. No apparent reason can be found for USPC introducing new brands at this time as they were producing many successful brands of their own, as well as brands retained from their acquisitions of National, Standard, NYCC, Dougherty, etc.

US35 AUCTION BRIDGE #708

USPC - RM Fact. This was a narrow version of the Jumbo index cards which originated about 1920 and was made for the relatively short time that Auction Bridge remained in vogue.

US36 WE-THEY #868

USPC, c1930. This was a popular narrow brand made during the surge in popularity in Contract Bridge. The deck was sold in pairs in a double box. One back was printed 'We' and the other 'They'. It was also made with a USPC Ace of Spades.

US33a

US33a HELMET #19

USPC, c1910. The identical deck with the Ace of Spades changed, removing the company name and the USPC trade mark from the pip. The Joker is the same, but the number is '19' and the word Joker replaces the 'US' in the corners. This brand appears to be have been made in an attempt to conceal the maker.

US36

Some of the brands mentioned in this chapter were also made for large customers as private brands. Decks from the regular line were used, substituting a special Ace of Spades and Joker. Large chains often sold decks under their own label. Examination of the face cards will usually reveal the brand and maker. USPC from about 1905 to 1940, made literally hundreds of special brands for specific stores, stationers, etc. and it is beyond the scope of this Encyclopedia to attempt to list them all. An example, shown here, by Personal Stationery Co., was easy to identify, as they retained the 'WE-THEY'.

Collectors might notice the absence of other well-known USPC brands (e.g. #500). These are listed in the National Card Co. chapter, as all brands that originated under their banner are listed there, even though USPC is the only name shown on the Ace of Spades. This was done to preserve continuity and to avoid showing the same brand in two places.

USPC manufactured three brands of Miniature or Patience size Playing Cards. In their advertisements they were often referred to as 'Toy' cards. Cards of this type enjoyed great favor in Europe and were popular in the United States in the 1890s and early 1900s. Thereafter they lost much of their appeal, with the result that examples of the brands listed here are less plentiful after, say, 1920. USPC did an excellent job of meeting the demand, in contrast to Andrew Dougherty, who, in their entire existence, never produced a miniature deck.

THIS is the first *legitimate* miniature playing card that has ever been produced by a playing card manufacturer; it has therefore all the qualities of the regular playing card. It is a *vest pocket card*, and useful for *Solitaire* and other games as well as a pleasing toy card for children.

EXTRA AD CARD IN CADETS

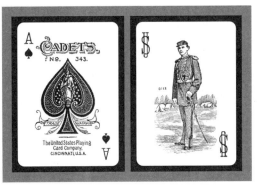

US37a

US37a CADETS #343
USPC, c1895. The second Ace of Spades was issued before 1900 and the Joker had the familiar 'US' placed in the corners.

US37b

US37b CADETS #343
USPC, c1910. The last Ace of Spades, in the simpler style common after 1910, was issued about that time.

US37

US37 CADETS #343
RMP, 1885. Cadets #343 was the first brand of miniatures to be produced by the company. There were several different Aces of Spades through the years, but only one style of Joker. The brand was phased out about 1925. The size of the card was 1 5/8 x 2 1/2 inches.

US38

US38 FAUNTLEROY #29
RMP, c1885. Shortly after the introduction of Cadets, an enameled edition of the same size was presented. Besides the Fauntleroy Ace of Spades and the multi-colored Fauntleroy Joker, there were special backs. Four of these (The Earl, Dick, Dearest and Mr. Hobbs) had a monochrome scene from the novel 'Little Lord Fauntleroy'.

US38a FAUNTLEROY #29
USPC, c1895. The new Ace of Spades was similar with just the change of the company name. The early editions of this deck kept the original Joker, but in later editions there was a new Joker, a mounted Lord Fauntleroy that lasted for the remaining years of issue.

US38a

US38b

US38b FAUNTLEROY #29
USPC, c1915. A new Ace of Spades with no sketch of Fauntleroy that lasted until the brand was phased out after World War II. Tastes in reading obviously changed, and youth of the later generations knew less of the classics and geometric back patterns replaced the special backs.

US39

US39 LITTLE DUKE #24
USPC, c1898. The smallest of all USPC miniatures. The courts are very attractive, being reminiscent of the standard French cards. The early issues came in a very collectible tin box with advertising for USPC and depictions of some of their brands.

US39a LITTLE DUKE #24 *(not pictured)*
USPC, c1910. A slightly later Ace for this brand which was phased out about 1950. The size of the Little Duke deck was 1 1/4 x 1 3/4 inches.

US40

US40 FORTUNE #814
USPC, c1918. A scarce deck that has nothing to do with fortune telling. We might assume the deck was made for export overseas due to the way it says "Cincinnati, United States of America" and comes with French style courts.

US40a FORTUNE #814 *(not pictured)*
USPC, c1920. The foreign version with four corner indices soon followed. In a 1923 price list, only this version was available.

US41 No.87 BEZIQUE
USPC - RM Fact., c1899. This was a slightly smaller deck than normal (2 1/4 x 3 1/4 inches) and came with a plain back in two series, A & B. Other than Faro #366, it is the only deck issued by USPC without indices.

US41

US42
USPC, c1920. A special Ace of Spades found on a deck made for a Steamship company. Probably this Ace was only used on special advertising decks.

US42

US43 BOXER *(not pictured)*
(now listed as PU13).

US44

US44 OXFORD
USPC, c1920. A special unidentified narrow brand.

US45 INDIAN (not pictured)
USPC, c1894. The box is marked with the name of the brand although the only clue on the deck is the 'Indian' Joker. The corner initials on the Joker leave no doubt as to the manufacturer of this deck.

US45a

US45a INDIAN
National PCC, c1890. A version with the National 'stars' in the corners of the Joker. This was probably a National brand before the take-over by USPC in 1894.

US46

US46 HORNET #6
(formerly MSW87), USPC, c1905. An inexpensive deck (15 cents per the box) made by USPC to compete anonymously with other inexpensive cards.

US46a

US46a HORNET #6
USPC, c1915. A later version with a generic Joker, similar to the one used in Canada where the deck was much more common and identified as being by USPC.

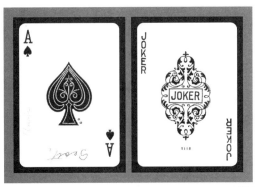

US47

US47 VOGUE #831
USPC, c1910. Another brand, with wide monotone named backs, made to compete anonymously with other makers' cheaper cards. It is interesting that both Hornet and Vogue were made in the Canadian plant with Aces showing the USPC name (see Chapter 16). It has a monotone pictorial back.

US48

US48 OWL & SPIDER
(formerly MSW105), USPC, c1927. This deck, previously unidentified, was made for the Bohemian Club in San Francisco.

The following four decks were originally thought to have been made by a company owned and operated by Piatnik of Vienna. Certainly Piatnik had an import company using that name after WWII, but it is unlikely that they were making cards for North America in the first/second decades of the 1900s. More importantly, close examination of the Diavolo cards (US51), and especially the courts, shows them to have been manufactured by USPC. It is highly likely, but not proven conclusively, that Atlantic Playing Card Co. of New York was a subsidiary or division of USPC, and, accordingly the decks have been moved to this Chapter.

US52

US52 MONOGRAM
Atlantic PCC, c1937. A narrow deck apparently made by this company, although 25 years later than the first three, and clearly made by USPC.

US49

US49 STUDIO
(formerly MSW29), Atlantic Playing Card Co., New York, c1910. The Ace is printed mainly in blue and it comes with monotone backs.

US50

US50 DOUBLE PINOCHLE
(formerly MSW30), Atlantic PCC, c1910. This deck only had the name on the box. Another deck identical to this one was issued by the Progressive Card Co. of New York and probably made by Russell, adding credence to the belief that Atlantic was part of the USPC/Russell group.

US53

US53 TOWER MFG. & NOVELTY CO.
(formerly MSW133), New York, 1905. A deck manufactured for this company by USPC.

US51

US51 DIAVOLO
(formerly MSW31), Atlantic PCC, c1910. This deck came with a great Joker, printed entirely in red!

US54

US54 SENATE
USPC, c1930. The Ace and Joker of a recently discovered wide brand of cards from about 1930 are shown below. They came in an interesting new style of tuckcase.

US54a

US54a WHIST
USPC, c1935. A narrow deck of high quality with a similar Ace.

US56

US56 INITIAL PERIOD
USPC, c1922. A private brand of narrow cards made for Chas. A. Stevens & Bros. of Chicago. Many other narrow, private brands were undoubtedly made by USPC during the 1920's and 1930's.

US55

US55 UNITED STATES PRINTING CO.
USPn., c1892. A scarce early advertising deck using the USPC trademark Ace and advertising their own company!

US57

US57 TRIUMPH
USPC, 1940. A wide deck of standard cards likely made for sale in the armed forces facilities.

US55a

US55a GLENLIVET WHISKY
USPC, c1905. We show another early advertising deck with the unique use of the USPC trademark Ace. It was very unusual for the Company to use their trademark when promoting another company's product.

US58

US58 MOHAWK
USPC, 1940. Another deck from this era, for some reason, listed on the box as being made by NYCC. The Ace and Joker were used commonly by USPC for other brands from the 1940's on.

US59 IEASE #801 *(not pictured)*
USPC - RM Fact., 1912. This deck with a 1912 tax stamp was made of a special substance giving it a non-glare appearance that was easy on the eyes! The deck is otherwise a standard Bicycle deck, US8c.

EARLY US AND NATIONAL ADVERTISMENT

THE NATIONAL CARD CO., INDIANAPOLIS AND NEW YORK

Samuel J. Murray, a printing pressman by trade, went to England as a young man to work in the playing card manufacturing facilities of Charles Goodall. In 1881, Robert Morgan convinced him to move to Cincinnati where Murray became a key employee of the Russell & Morgan Company. His employment marked the rapid advancement of Russell & Morgan to the first rank as a manufacturer of playing cards, and one of the most efficient. Murray was very inventive and designed and installed much of the machinery used in the manufacture. One invention, an automatic punch machine, increased output fourfold and reduced labor costs by 65%. In 1886 he left Russell & Morgan because of a disagreement with his associates about his advancement.

Jobless, he moved to Indianapolis and established The National Card Company. This new company became such a serious menace to Russell & Morgan, that they bought it out on March 17, 1893. As part of the arrangement, Mr. Murray was given a large block of stock and put in charge of all playing card manufacturing at USPC. In 1894 National was merged with the United States Printing Co. as the key component in the formation of the United States Playing Card Co.

Over the whole period of its existence, National produced a relatively small number of brands, and most of these remained in production from its start until well into the 1920s. Most of the National brands in production at the time of the merger were retained and marketed by USPC. Even today, a few of the old National brands, including Rambler and Apollo, are still in production for certain markets.

A few new brands were added. For example, the first 500 deck, called Full House, was patented by USPC for National in 1896. Most of the brands kept the original National Ace of Spades for many years, although some Aces were changed to read 'National Card Co., Cincinnati.', some added 'USPC' to the original National Ace, and some were completely changed and switched to the USPC line. The National factory in Indianapolis was operated successfully by USPC until about 1900.

NU1

NU1 NATIONAL STEAMBOATS #9
National Card Co. (NCC), c1882. All companies producing playing cards had their own version of the Steamboat brand, usually the least expensive cards in the line. The colorful Watermelon Joker was used for both NU1 and NU2.

NU2

NU2 SUPERIOR STEAMBOATS #9
c1885. This deck was of slightly better quality and soon replaced NU1. It was retained by USPC until about 1910.

NU3

NU3 STEAMBOAT #9
c1910. Note the 'Star in Circle' indices on the Joker. This is a positive way to identify cards by National. This deck replaced NU1 and NU2 and was used as the new "low price leader". The Steamboat Joker, similar to the Joker of US7, had 'Star in Circle' indices.

NU4

NU6

NU4 OWLS

c1885. This was one of the earliest brands by National and the only one known to have been dropped prior to the USPC merger.

NU4a

NU4a OWLS

c1882. The same deck as NU4, but either made as a private brand, or as an earlier version prior to the adoption of the company name. The Joker also omitted the National identification.

Superior Steamboats #9 (NU2) and the decks which follow were all offered by USPC in their 1896 price list and catalogue. The decks are not listed chronologically, but rather by quality and price. NU2 is the least expensive, while National Club (NU19) is at the top of the line.

NU5

NU5 ARROWS #11

c1885. The Arrow brand was listed as the brand just above Steamboats. They were adapted from the plain National cards (NU5a), originally sold as Arrows. The Arrows brand was continued until about 1905.

NU5a NATIONAL ARROWS (not pictured)

c1882. This was a simple version of NU5 without the name on the Ace of Spades. It used the same Joker.

NU6 ALADDIN #1001

c1885. This brand was the next step up from Arrows. They were unenameled with a "process finish" and a hard surface.

NU6a ALADDIN #1001 (not pictured)

by NCC, Cincinnati, c1902. These were made after the sale of the Indiana plant. There was, as yet, no apparent effort to phase out the National name. The Ace was virtually the same as NU6 and the Joker was like NU6b.

NU6b

NU6b ALADDIN #1001

By NCC (USPC) c1910. USPC was placed on the Ace of Spades in addition to the National name. No records exist as to the date when the National name fell from general use. Indeed, some brands of National cards are still being manufactured today.

NU6c ALADDIN #1004 (not pictured)

(USPC), 1928. This is an unusual deck as it has no indices. Other than that, the Ace of Spades and Joker are virtually the same as NU6b and the deck may have been made to satisfy the tastes of the die-hards who still refused to accept decks with indices.

The following picture is of the National Card Co. plant as it looked in 1891. The plant was located at 5th and Eggleston in Indianapolis. At that time National also maintained an Eastern office at 221-227 Canal St. in New York City.

NATIONAL FACTORY

NU7b

NU7b RAMBLER #22
c1895. This and subsequent Rambler Aces of Spades had the name on the Ace.

NU7c RAMBLER #22 *(not pictured)*
c1905. Cincinnati replaced Indianapolis on the Ace of Spades. The alterations were made on the new plates after the sale of the Indiana plant.

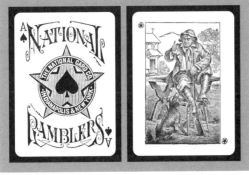

NU7

NU7d

NU7d RAMBLER #22
(USPC), c1910. USPC added to the Ace of Spades.

NU7 RAMBLER #22
c1885. This deck was next in line to Aladdin. It also featured the "hard surface, process finish", but was advertised as "Unequaled in perfect slip". Rambler was one of their most popular brands, and is still made in limited quantities today. Notice that the central design on the first Ace is the same as that of the Arrows NU5.

All National brands could be purchased with gold edges. The brand numbers, on the box only, were always one higher on gold edged decks - Steamboat #10, Rambler #23, etc.

NU7a

NU7a RAMBLER #22
c1890. This newly designed Ace of Spades was to run the life of the issue, with changes only in the company name and addresses. As often happened with early cards, there was no brand name on the Ace of Spades, only on the box. The pictured Joker was used for this and subsequent versions.

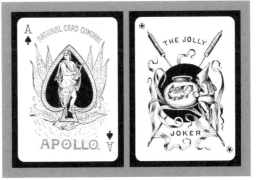

NU8

NU8 APOLLO #33
c1890. A higher quality deck advertised as "Highly enameled, unequaled for high finish, elasticity, and durability." It featured a Fishbowl Joker.

NU8-1

NU8-1 APOLLO #33

c1882. We recently discovered that Apollo also had a very early version and have pictured the 'Steamboat' quality Ace of Spades. Unfortunately, we have not seen a copy of the Joker.

NU9a

NU9a PINOCHLE #300

(USPC), c1905. The USPC edition for this popular game.

NU8a

NU8a APOLLO #33

c1895. The same brand, but with a completely redesigned Ace of Spades and a new Apollo Joker. The flags were removed from the Ace.

NU10

NU10 COLUMBIA #133

c1895. Enameled, aluminum surface, made expressly for whist. NU8, NU9 and NU10 were the same quality cards, but were intended for different games, with different purposes. They were advertised as "French size - 2 1/4 x 3 1/2 inches".

NU8b

NU8b APOLLO #33

(USPC), c1905. Since 1900, "Aluminum surface" was featured on the Apollo brand. The same Joker was used.

NU10a COLUMBIA #133

(USPC), c1900. The same deck as NU10, with the addition of the words 'Whist Cards' to the Ace of Spades. The same Joker was used.

NU10a

NU9

NU9 PINOCHLE #300

c1895. Enameled Pinochle cards, put up in a 48 card pack. This brand featured a beautifully engraved Ace of Spades.

NU10b

NU10b COLUMBIA #133

(USPC), c1905. USPC was added to both the Ace and Joker.

NU10c

NU10c COLUMBIA #133

(USPC), c1920. National Card Company no longer appeared on the Ace of Spades and 'Bridge' replaces 'Whist Cards' in the center of the spade pip.

NU11

NU11 CRESCENT #44

c1885. Double enameled on specially processed linen stock. The decks have back designs featuring household pets. At the turn of the century these were changed to artistic landscapes.

NU12

NU12 BOSTON #55

c1885. Early advertisement: "Fine linen stock, double enamel, highest finish and slip surface. New and beautiful backs, designed by leading artists." Clearly one of their highest quality decks!

NU12a

NU 12a BOSTON #55

SERIES 'A', c1896. A later edition of NU12 with the same Joker issued as a 'Fashion Series' after the sale to USPC.

NU13

NU13 FULL HOUSE POKER CARDS #555

Patented June 30, 1896. This was a 60 card deck plus a Joker including the regular 52 cards plus 11 and 12 spot pips. It is the first example of a deck carried through to the USPC line, using the same patent. USPC called the deck "500", but examples of it will still be listed here, under NU13. This Ace was almost identical to NU12, the only variation being the name. There are two variations of the Ace of Spades, the one pictured here says 'Patent Applied For' and the second gives the patent date.

NU13a   *NU13b*

NU13a 500

USPC RM Factories, c1910 (with 11 & 12 spots), for Poker, 500, Rum, etc. Notice the new Joker, with Five Hundred substituted for National Card Co., and the 'US' monogram replacing the 'Star in Circle' indices. The Ace is like NU13b without the words "With 11 and 12 spots".

NU13b 500

USPC RM Factories, c1910. The same Joker was used as in NU13a.

NU13c 500 (not pictured)

USPC, c1925. Now a staple in the USPC line, this deck was increased to a 62 card deck with the addition of two 13 spots.

NU13d

NU13d 500
USPC, Russell & Morgan Factories, c1905. An earlier USPC version has turned up with both a different Ace of Spades and Joker.

NU16

NU16 LENOX #67
c1892. Gold edges, gold backs, double enameled, and linen stock made this deck one of the most expensive in their catalogue. The name Lenox did not appear on this Ace and the Joker was identical to NU13.

NU14

NU14 TENNIS #144
c1885. Double enameled, on specially prepared linen stock, this was one of the earliest Whist size decks made by any manufacturer. Due to the 19th century dress of the tennis players on the Ace, this brand was dropped around 1900 and replaced by the Columbia brand (see NU10). The same Joker was used as in Columbia, but with the 'Tennis' name sometimes added inside the moon.

NU16a

NU16a LENOX #67
by USPC RM Factories, c1905. A case of the brand being converted to the USPC line with no mention of National.

NU17

NU17 EL DORADO #49
c1885. "Gold edges, double enameled, linen stock, beautiful backs, each in many colors". A scarce, luxury card indicating that it was not a good seller and likely discontinued early. It comes with a beautiful Joker, very similar to that of Crescent #44.

NU15

NU15 NATIONAL WHIST #175
c1890. Double enameled, linen stock Whist cards. The cards are the same quality and have the same style Ace of Spades as the "Top of the line" National Club NU19.

NU15a NATIONAL WHIST #175 *(not pictured)*
(USPC), c1905. USPC was added to the Ace of Spades and Joker.

NU18

NU18 BIJOU #1
c1885. A special deck, which when introduced was the highest priced card in the line. The first ad said: "A perfect gem, the most elaborate playing card in the market. Beautiful designs printed in six colors. Put up in an imported cloth case or an elegant leather case". The NU15 Joker was used.

NU18a

NU18b

NU18a BIJOU #1

c1895. The Ace of Spades was modified and incorporated the #1 inside the spade pip and a new Joker was designed.

NU18b BIJOU #1

USPC, RM Factories, c1905. USPC adopted this brand early as their top Whist card. It ran concurrently with the NCC version for some time. The deck featured the NU18a Joker with 'USPC' in the corners.

NU19

NU19a

NU19 NATIONAL CLUB #75

c1885. The top of the National line was advertised as follows: "The best card made. Pure linen stock and double enameled. Unrivaled in all essential features constituting a perfect Playing Card. Made expressly for club use".

NU19a NATIONAL CLUB #752

c1900. The '2' added to the regular digit, denotes this deck as a special order. Clubs, placing orders for large quantities, could have special monogrammed backs with their own name or logo. It used the same Joker.

NU20

NU20 AMERICAN WHIST LEAGUE

An example of a special issue, made in large quantities, for a volume user. This deck was made for the American Whist League in 1897 and was made on National Club Seconds (note the grade 'B'). It is curious that the AWL did not order whist size cards! This deck is also classified as BW8 in the Bridge and Whist Chapter.

Finally, it should be noted that National Card Co. made special advertising decks both before and after they were acquired by USPC in 1894. Few examples have been positively identified as National decks because the manufacturer's name was usually not placed on the cards or box when advertising another company's product. In these cases, however, careful examination of the courts, Joker, and box will often reveal the maker. We have shown one such early advertising deck for Jos. Schlitz Brew'g Co. for illustrative purposes. The courts tell the story, but the Joker depicted here confirms it is a National deck. Note the star in the circle in the corners and the three 'Brownies', so typical of National jokers, enjoying the product.

NUAd

EL DORADO #49 WRAPPER

LENOX #67 WRAPPER

PERFECTION PLAYING CARD COMPANY

Perfection Playing Card Co. of New York and Perfection Playing Card Co. of Philadelphia each produced several editions of playing cards, which were almost identical. While most of the examples of cards manufactured by Perfection that are found today are standard cards, they were known to make advertising decks with both generic and special Aces of Spades and Jokers. In addition, they manufactured railway decks for the Rock Island Line.

Playing Cards manufactured by Perfection of New York have been found with evidence dating them as late as 1915, but no Philadelphia cards have been found dating after 1890. According to a notation in the records of the United States Playing Card Company, Perfection was established in 1885 in Philadelphia and moved to New York in 1889 or 1890.

It was long thought, because no card, box or advertisement had been discovered which had reference to Perfection Playing Card Co. of both New York and Philadelphia, that they started into business in Philadelphia and subsequently moved to New York. However, we now know that from 1886 to at least 1888, Perfection Playing Card Co. of Philadelphia was located at 819-821 Filbert Street and had an office at 336 Broadway. In addition, we know that Perfection Playing Card Co. of New York had a factory and offices at 100-102 Reade Street, New York sometime after this, and certainly before 1900.

1896 LETTERHEAD

We have a copy of a notice dated April 7, 1888, whereby Henry M. Rosenbaum withdrew from the business with it being continued by Edward Stern and Charles Dittman. On July 20, 1894, the Perfection Playing Card Co. of New York, under the management of Messrs. Charles Dittman, Bertram Grossbeck and George F. Jones, agreed to sell the business to a group consisting of USPC (50%), NYCC (25%) and Andrew Dougherty (25%). Their desire was to keep Perfection intact and conducted as a seperate business, under the control of the three firms which contributed to its purchase. As USPC eventually absorbed all of the companies participating in the purchase of Perfection,

we can sum up by saying "Perfection was absorbed by USPC in 1894."

EARLY STORE CARD

An 1886 price list for Perfection listed the following brands with index guides, round corners and every pack in a neat box. The brands from #350 to #700 came in both plain and gilt edges:

Roosters #100	Favorites #300
Tip-Top #350	Champion #400
Jockey Clubs #500	Cube Backs #550
Specials #700	Solo, Euchre and Pinochle

Later they added these brands which were on their 1893 price list: Steamboats #90, Aurora #475, Coronet #525, Double Pinochle, Daisy, Premium and Victoria. By 1900 more brands were listed, including Eire #57, Meteor #150, Winner #333, Leader #325, Monarch #365 (French size), Cameo, Crest and Elite.

Except for certain issues of Champion #400 (PU7), the Perfection Aces of Spades did not show the brand name, and therefore Perfection brands can only be identified from the box or wrapper.

Perfection also made cards sold under the names of several distribution companies. These included:

The New Orleans Playing Card Company, New Orleans, La.
The Anchor Playing Card Company, New Orleans, La.
Regal Playing Cards
Silver City Playing Cards
Diamond Card Co., New York
Valley City Playing Cards

The Aces of Spades listed below, from PU1 to PU5 and PU8 to PU11, were likely each used for many different brands. In addition, we have found that Perfection used the various Jokers interchangeably with different Aces. We have also found that there are several variations of certain Jokers. For this reason, while we have shown the Jokers as belonging with certain Aces, one should not be surprised to find that their Perfection deck has a Joker from a different listing.

PU1

PU1 TIP-TOP #350

Philadelphia, c1885. Perhaps the earliest deck from Philadelphia. While we have seen this Ace and Joker in a Tip-Top box, they were likely issued with each brand.

PU3

PU3 STEAMBOATS #90

Philadelphia, c1885. The identical Ace of Spades was used as in PU2, but the Joker and box were different.

PU1a

PU1a TIP-TOP #350

Philadelphia, c1886. This Ace is slightly different. Three different versions of the Joker are known, one in black and white, one in black, white and red and one in full color also used for this deck.

PU4

PU4 LEADER #325

New York, c1890. The deck is similar in every respect to PU2, but with the New York Ace of Spades. There were at least four different Jokers used with this deck, the two pictured and two versions of the Jumping Jack pictured with PU2, a black and white and one printed entirely in red. We have also seen a copy of this deck with advertising for a building contractor overprinted on the backs.

PU5 STEAMBOATS #90 *(not pictured)*

New York, c1888. The same Ace of Spades as PU4. It came with at least two different Jokers, one the same as PU3 and the Rooster Joker of PU5a.

PU2

PU2 TIP-TOP #350

Philadelphia, c1887. The cards in this deck look slightly later than PU1. It came with at least three different Jokers. We have found several Perfection decks in special boxes for R. H. Stearns & Co., Engravers and Stationers, Boston, with the Perfection brand name and number on the ends of the boxes.

PU5a

PU5a ROOSTERS #100

Philadelphia, c1885. The PU2 Ace in a Roosters #100 box with the Rooster Joker.

PU6

PU9

PU6 GEOGRAPHICAL EUCHRE
Philadelphia, 1886. This is an unusual deck named Euchre, but containing all 52 cards (PU2 Ace) and the rooster Joker. The cards each have geographical information about the United States printed on all four sides. There was a special game with instructions included in the original box.

PU9 PERFECTION
London, c1890. It appears Perfection had some involvement with the English market for playing cards. The deck shown has London as the prominent address on the Ace of Spades, along with Paris and New York, and was probably manufactured in the USA for export.

PU7 CHAMPION #400
Philadelphia, c.1890. The only brand by Perfection with the brand name printed on the Ace of Spades. All other decks by Perfection had their names and numbers printed on the boxes only. This brand used the PU1 Joker.

PU7

PU10

PU10 PERFECTION
New York, c1890. The same Ace of Spades as PU1 but New York replaces Philadelphia. The indices, printing, etc. date the New York version somewhat later.

PU8 MONARCH #365 *(not pictured)*
New York, c1895. The only narrow Perfection deck found to date was advertised as being "French size" for Whist players. It has the same Ace as PU8a.

PU11   PU12

PU8a PERFECTION #450
New York, c1892. A wide version of the PU8 Ace of Spades was also produced, likely earlier.

PU11 CHAMPION #400
Philadelphia, c1885. An Ace of Spades that is different from the early wide Philadelphia decks, but almost identical to the Ace used in PU8a. This deck came with the PU1 Joker.

PU12 COLUMBUS
New York, 1893. This deck was issued for the Columbian Exposition in 1893 and it has also been included in the Exposition category as SX38. The top of the box has "1492", the bottom "1893" and the Joker is a portrait of Columbus. It is of interest in this section due to the use of an identical Ace of Spades to that of PU8a.

PU8a

PU13

PU13 BOXER

(formerly US43), New York, c1895. This deck does not have any maker's name but the courts, which are identical to those of PU15, prove it was made by Perfection. We expect it is somehow related to the Boxer rebellion in the 1890s as the Joker shows a Chinese dragon and the backs are China related.

PU14

PU14 ERIE #57

New York, c1895. This brand, with the same courts as PU13 and PU15, was listed in Perfection's 1900 price list. Clearly the deck had something to do with the Erie Canal.

PU15   *PU16*

PU15 ROOSTER #38

Excelsior Playing Card Co., c1885. This deck was obviously made by Perfection but only mentions Excelsior Playing Card Co. on the Ace of Spades. It is essentially the same deck as PU5a except for the name on the Ace and the brand number on the box. It used the same 'Rooster' Joker.

PU16 PERFECTION

Chicago, c1895. This Ace of Spades is from a sample at the USPC Collection and was likely made for a local distributor in Chicago, similar to those for the distribution companies.

PU17   *PU17a*

PU17 DIAMOND CARD CO.

New York, c1895. Another special Ace made for a distribution company, in this case Diamond Card Co. They made the following brands: Railroads #110, Bengals #120, Sultanas #130, The Sun #140 and Pacifics #150.

PU17a DIAMOND CARD CO.

New York, c1895. Another Ace used by this company.

PU18 MANHATTAN CARD CO.

New York, c1900. The courts on this deck are slightly different than other Perfection decks, but the Ace of Spades makes us believe it was made by Perfection.

PU18

PU19 VALLEY CITY PLAYING CARDS

c1897. Another deck suspected to be by Perfection.

PU19

Several advertising decks were manufactured by Perfection, notably American Electrical, advertising electric wires, with the PU4 Ace of Spades and a special Joker, and Acorn Stoves, with a special Ace of Spades and the PU5a Rooster Joker.

PYRAMID PLAYING CARD COMPANY

The Pyramid Playing Card Company was incorporated in May of 1920. The original stockholders were Abdala Barsa, Simon Barsa and Suad Salfeety and they, along with Gibran Ateyeh and Michael Kayata, were the first directors. The name Pyramid might have stemmed from their apparent Egyptian origins. Abdala Barsa was the first President.

The company commenced operations in Brooklyn, New York from premises at 351-353 Jay St. They produced many different brands, and must have been reasonably successful judging by the number of examples of their products that can be found today.

The company, like so many of the others, soon attracted the attention of the United States Playing Card Co. In May of 1924, with the exception of the Barsas, the directors resigned and were replaced by directors active in USPC and Russell PCC.

In February 1925, Messrs. John Omwake, Arthur Morgan and C.E. Albert, all officers of USPC, held 2,836 shares of the capital stock of Pyramid. They voted that the company be continued under the leadership of NYCC, then in the USPC fold.

In February 1927, the company leased new premises on Livingston Street in Brooklyn. At this time the last interests of the originating founders of Pyramid ceased to exist. In May 1929, the Brooklyn plant was dismantled and the offices moved to New York (an extra card in one deck shows the address as 153-157 West 23rd Street). The corporation was finally dissolved in 1933.

PY2 PYRAMID SPECIAL
(Formerly MSW112), Brooklyn, c1920. A different Ace for this brand which utilized the standard Pyramid Joker.

PY2

PY3 MUTUAL PLAYING CARDS
(Formerly MSW113), Brooklyn, c1922. This brand was issued in both regular and Pinochle cards, again with the Pyramid Joker.

PY3

PY4 HOME RUN
(Formerly MSW114), Brooklyn, c1922. What better brand name for a firm making cards in the shadow of Ebbetts Field, home of the then Brooklyn Dodgers? The brand utilized a nice baseball Ace and, unfortunately, the standard Joker. It also came in a plainer version without the brand or company name.

PY4

PY1

PY1 WINNER
Pyramid Playing Card Co., Brooklyn, c1920. An interesting Ace of Spades, in red, white and black, with the standard Pyramid Joker used, with minor variations, for most of their standard wide decks.

PY5 IRIS PLAYING CARD
(Formerly MSW115), Brooklyn, c1920. This is the only Ace of Spades to carry the address of the company. It was issued as a 48 card Pinochle deck.

PY5

PY6

PY6 PYRAMID #100
(Formerly MSW116), New York, c1926. Beautiful new styled court cards were featured in this four way indices deck. Once again the standard Joker was issued with the deck.

PY6a PYRAMID #100 (not pictured)
New York, c1926. The same beautiful courts in a version with the normal two indices.

PY6b PYRAMID #100 (not pictured)
New York, c1926. This deck has the same Ace of Spades but uses the regular courts.

PY7

PY7 BLUE NILE
(Formerly MSW117), Brooklyn, c1923. The first of their narrow decks, again with a version of the same Joker.

PY7a

PY7a PETER PAN
(formerly MSW122a), Brooklyn, c1924. A similar Ace of Spades with Peter Pan identifying the brand on both the Ace and the Joker.

PY8

PY8 BLUE STAR
(Formerly MSW118), Brooklyn, c1924. Another narrow deck with the Pyramid Joker which also came in a plain edition without the brand name.

PY8a

PY8a PRINCESS
Brooklyn, c1924. This recently discovered brand has a "P" inside the blue star. The common Joker has a quotation underneath the base "Forty Centuries Look Down Upon You".

PY9

PY9 UMPIRE
(formerly MSW119), New York, c1926. This narrow deck came in two versions – one with and one without the Umpire name on the Ace. Both versions used the special Umpire Joker (PY9a).

PY9a

PY9a UMPIRE
New York, 1927. This deck had the pyramid in the center of the Spade Ace and came with USPC courts. Not only that, the cellophane wrapper has a Dougherty tax stamp. Clearly it was made during the integration of Pyramid into USPC and its affiliates.

PY10

PY10 A.A.O.N.M.S

(Formerly MSW120) Pyramid PCC, Brooklyn, 1923. Like most of the other manufacturers, Pyramid made special cards. This deck was made for the Kismet Temple, a branch of the Shriners. It is a handsome deck, beautifully executed.

PY11 SUNBEAM *(not pictured)*
(Formerly MSW121), Pyramid PCC, New York, c1930. This deck has no name on the Ace of Spades but is fully identified on the box and has the normal Pyramid Joker.

PY12

PY12 BARSA B
Pyramid PCC, Brooklyn, c1920. Perhaps their first deck, as it was made on Jay Street and was named after the President and Vice-President. It includes an unusual colored Joker.

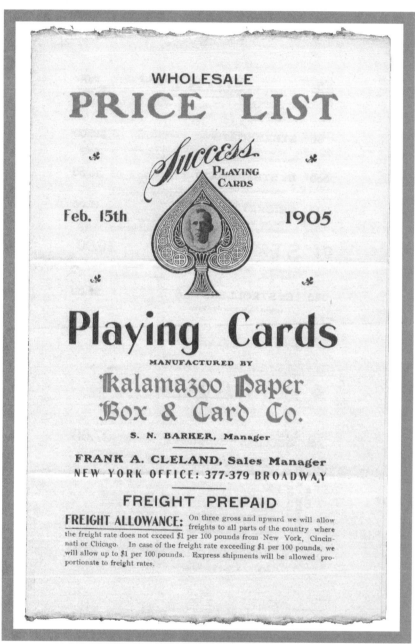

KPBC PRICE LIST COVER

WILLIS W. RUSSELL, KALAMAZOO, RUSSELL, AMERICAN BANK NOTE, AND OTHER RELATED MAKERS

In the early part of the 20th century, Willis W. Russell, no relation of A. O. Russell, founder of Russell & Morgan Co., started a modest playing card business in Milltown, a small town in northern New Jersey. Production commenced in 1905 and from the beginning the firm made a fine product and introduced numerous brands that became well known to the card playing public. Mr. Russell's pride in his merchandise was evident, as his portrait and signature appeared on the Ace of Spades of every Willis W. Russell deck produced.

COLORED AD FOR RUSSELL

RU1

RU1 RUSSELL'S RECRUITS
Willis W. Russell, Milltown, N.J., 1906. It is interesting that the cards claimed 'Better than Steamboats' on the Ace of Spades and box of every deck. Recruits was clearly taking aim at 'Steamboats', normally the cheapest line produced by other makers. (Recruits had the 'long distance pips' that became a popular feature on some of the Russell decks). On these cards the Spades and Hearts had shading in the center, but the Clubs and Diamonds did not.

The following quotation from an extra card inserted in a 1906 Russell deck explains: "Card enthusiasts use Russell's long distance faces with satisfaction. They are a practical solution to the frequent confusion of the suits of same colors, particularly in Bridge where one hand is always played at long distance, also in Pinochle, Poker, Cribbage, etc. Our improvement prevents this by tinting, or ornamenting, one of the suits of each color, leaving the opposite suit solid as shown here."

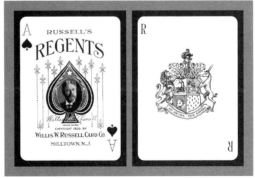

RU2

RU1a RECRUITS *(not pictured)*
Russell PCC, Milltown, N.J., c1912. This edition, with the RU18 Ace of Spades and the RU1 Joker, was published just after the purchase by Rosenthal.

RU2 RUSSELL'S REGENTS
Willis W. Russell, Milltown, N.J., 1906. Regents was the first Russell deck to use the 'long distance pips' concept. Regents were described on the box as 'Art Series – Gilt Edged' and had named pictorial backs.

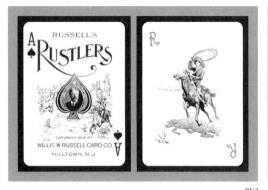

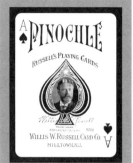

RU3   RU4

RU3 RUSSELL'S RUSTLERS

Willis W. Russell, Milltown, N.J., 1906 (refer also to SE16, a Rustlers deck with a baseball motif). The standard Rustlers had backs with a cowboy motif. They were also produced with gilt edges.

RU4 RUSSELL'S PINOCHLE

Willis W. Russell, Milltown, N.J., 1906. Perhaps there was no way to start Pinochle with an 'R'! This was a 48 card deck without a Joker. It had the long distance pips.

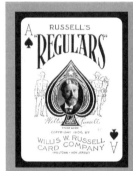

RU5

RU5 RUSSELL'S REGULARS

Willis W. Russell, Milltown, N.J., 1906. A brand aimed at the armed forces with beautiful named backs such as 'Sabre' and 'Liberty'. Regulars also had the long distance pips. The pictured box is one of the most attractive ever produced. It has a colorful front with the card back on the reverse.

RU5a RUSSELL'S REGULARS SKAT *(not pictured)*

Willis W. Russell, Milltown, N.J., 1907. Russell made a 32 card deck, in a special box, with the Regulars' Ace for the National Skat Congress in June 1907.

RU5b RATTLERS *(not pictured)*

Willis W. Russell, Milltown, N.J., 1906. A Russell's Regulars deck sold in a 'Rattlers' box with no identification on the box indicating it was made by Willis W. Russell.

RU6   RU6a   RU6b

RU6 RUSSELL'S MOGUL

Willis W. Russell, Milltown, N.J., 1906. A brand of standard cards that did not start with 'R'. They were made with regular pips only.

RU6a RUSSELL'S CLUB MOGUL

Willis W. Russell, Milltown, N.J., 1906. The Ace of Spades was different, but the Joker, box and quality were identical to RU6.

RU6b MOGUL #100

Russell PCC, New York, c1914. This is a later production using the RU18 Ace with a Joker reminiscent of the early version.

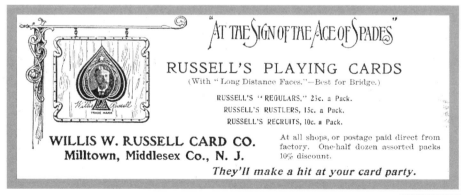

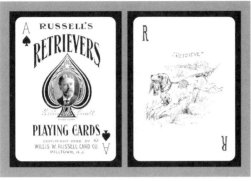

RU7

RU7 RUSSELL'S RETRIEVERS #40
Willis W. Russell, Milltown, N.J., 1908. The latest copyright date to appear on any Willis W. Russell card.

RU8

RU8 RUSSELL'S BLUE RIBBON #323
Willis W. Russell, Milltown, N.J., 1907. This deck also did not begin with 'R', and was another brand to bear a style number. It turned out to be the longest lasting of any of the brands and Blue Ribbon carries on today as a popular USPC brand. There are a large number of minor variations of this Ace and a few are shown below.

RU8a

RU8a RUSSELL'S BLUE RIBBON
Russell Playing Card Co., New York, c1930. The second edition of the Blue Ribbon Ace which sometimes came without the brand name (as shown here) and sometimes with it.

RU8b BLUE RIBBON *(not pictured)*
USPC, c1935. This Blue Ribbon Ace, RU8a but with the USPC name, is still being used by USPC today. It uses the RU8a Joker.

RU8c

RU8c SQUARED DEALERS #23
Willis W. Russell, Milltown, N.J., 1907. A deck with no indices and squared corners for use in dealing boxes. This type of deck was used primarily in Faro. The Blue Ribbon name was used but note the brand number was changed to '#23'. The RU8 Joker, without the 'R' in the corners was used.

RU8d

RU8d BLUE RIBBON
Russell Playing Card Co., New York, c1912. An earlier version of the narrow Blue Ribbon with the number '121' in the Joker.

EXTRA CARD

RU EXTRA CARD
An extra card provided in each Russell deck to allow the owner to circle a lost card and send to W.W. Russell for a replacement.

RU8e

RU8e ROULETTE CARDS
Willis W. Russell, Milltown, N.J., 1905. A gambling card game copyrighted and produced by Russell in 1905. Alonzo K. Ferris, who likely licensed production to Russell, also copyrighted the game in 1905. We suspect these were the first cards produced by Russell. They were sold with an attractive instructional booklet.

Despite producing a superior product, Willis W. Russell Card Co. was not making money and late in 1909, the firm filed for bankruptcy. Operations continued in the hands of receivers, but conditions continued to deteriorate and in 1911 they ceased operations and decided to sell the company at public auction.

We will continue with the discussion of Willis Russell's business a little later in this chapter. But before we do, we must move to a brief discussion of Benjamin Rosenthal. Rosenthal first entered the playing card industry as a young man of 19. He joined the American Playing Card Co., of Kalamazoo, Michigan as a salesman and for the first three months averaged a gross income of $10 per week! During the next few years he built up a lucrative business, mainly in the southern States. Unfortunately, his success created a problem and the manager of APCC soon advised him that the son of the President wanted his territory. Rosenthal lost his job and while understandably provoked, set about procuring another playing card connection.

The Kalamazoo Paper Box & Card Co., founded by S. N. Barker, had been in the playing card business for about a year, and Mr. Rosenthal applied for a selling agency, which he quickly secured. Shortly afterward he found out why - the company was practically bankrupt. Disliking failure, he set about reorganizing the line and the sales force and the business prospered. The increase was so pronounced that the company, with inadequate capital, found itself in even more difficulty. So Rosenthal turned his attention to reorganizing the company's finances, and eventually got it back on a financially sound basis. The Board of Directors, recognizing that the young man possessed unusual talents, rapidly promoted

him from sales manager to general manager, and ultimately to Vice President. He was also invited to join the Kalamazoo Paper Box & Card Co. Board of Directors.

The Kalamazoo Paper Box & Card Co. started producing playing cards in 1903. On June 11, 1906, the corporate name was changed to Kalamazoo Playing Card Co. The paper box business was disposed of and the plant was devoted to the manufacture of playing cards. George E. Bardeen, president of a large paper mill in Otsego, Michigan, was appointed President of the company. Benjamin Rosenthal was elected Vice President and Director of Sales, and Mr. Barker was retained as Manager.

RU9a

RU9a STEAMBOATS #666
Kalamazoo Paper Box and Card Co., c1903. The Ace is similar to RU9 but with the initials of the company in the center in place of the picture of the founder.

RU9

RU9 STEAMBOATS #666
Kalamazoo Paper Box and Card Co., c1903. Probably the least expensive card in the line. They also made a Steamboats #66. The picture in the center of the Ace of Spades is that of S.N. Barker, the founder of the company. It is odd that the only two early companies that utilized the portrait of their founder were soon to be united by circumstances that none could foresee. The Joker comes in two versions, one with the word 'Joker' and one without.

RU10

RU10 STROLLERS #04
Kalamazoo Paper Box and Card Co., c1903. The date of issue of any of KPB deck is an estimate, but as the name was used for only three years, it has to be fairly accurate. Another similarity between the Willis W. Russell Co. and KPB was the use of one letter in naming their brands – 'R' for Russell and 'S' for Kalamazoo.

RU10a

RU10a LILY STROLLERS #04

1904. In late 1904, KPB introduced their Lily brand quality with 'Swansdown' finish. For the first time the Ace of Spades did not show a man's picture. The Joker also changed slightly from the original Strollers (RU10) and the word 'Lily' appears on the flap of the box.

RU10b

RU10b STROLLERS #04

Kalamazoo Paper Box and Card Co., c1904. Another version of the this Ace with the initials replacing the picture in the spade pip.

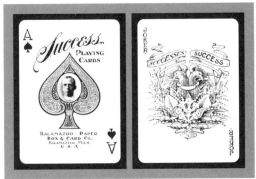

RU11

RU11 SUCCESS #28

Kalamazoo Paper Box and Card Co., c1903. KPB also made a Pinochle #48 deck with this Ace of Spades.

RU12

RU12 KALAMAZOO

Kalamazoo Playing Card Co., c1906. KPC used only two Aces of Spades during the life of the company. The brand names appeared only on the boxes and Jokers. This was a system used by Rosenthal's first employer, the American Playing Card Co. The brand shown here is the successor to Success #28 by KPB, but this Ace and the following one (RU13), were used on many other brands including, Aristocrat, Auction Pinochle #161, Beauty (with named backs), Bridge Whist #60, Chancellor Club #228, Chancellor Pinochle #164, Club Fellow #626, Cricket #13, Eureka #010, Fads and Fancies #150 (RU27), Hawk #0000, Idle Hour, Queen Quality (with named backs), Square Deal (RU28), Strollers, Torpedo #327 and Wauna. We have also shown the 'Cricket #13' and 'Chancellor #228' Jokers.

RU13

RU13 KALAMAZOO

Kalamazoo Playing Card Co., c1910. The second Ace used by KPC with Smart Set #400, Steamboat 7–11 and Idle Hour Jokers.

Both of the above Aces had brands that used backs with lovely monotone pictures. They are popular with back collectors, who favor wide named decks.

Returning to the story of Willis Russell, we find that the Willis W. Russell Card Company of Milltown, New Jersey, which had been in the hands of receivers for over two years, was to be sold at public auction. Mr. Rosenthal, building on his success at Kalamazoo, endeavored to enlist the interest of his Kalamazoo associates to purchase the Russell company. In the end they demurred and Rosenthal went to New Jersey to purchase it himself. He subsequently merged Willis W. Russell and Kalamazoo Playing Card Co. into the Russell Playing Card Company, moved everything to the Milltown plant and arranged to have himself appointed president of the new company.

Russell Playing Card Co. met with immediate success. Then, a year or so later, in early 1914, when the famed American Bank Note Company decided to discontinue their playing card line, Rosenthal acquired that business as well, and brought it to the Milltown plant. From that point on, Russell's growth was consistent and rapid. It quickly became one of the most prosperous industries in northern New Jersey and the second largest playing card maker in the United States manufacturing as many as 50,000 decks per day.

RU15

RU15 AMERICAN BANK NOTE
GERMAN SIZE, New York, c1910. This special longer and narrower size (2 3/8 x 3 3/4 inches) was issued for Pinochle (48 cards) and Double Pinochle #698 (64 cards) use. It was also made in a 53 card deck ('Club Edition') including the colorful Joker shown here.

RU16 AMERICAN BANK NOTE WHIST #454
New York, c1910. The third size deck made by ABN was this whist size 2 1/4 x 3 1/2 inches. Any of the several brands of these narrower cards will be grouped under this number. Some decks of this size were issued in an interesting series of Alice in Wonderland backs.

RU16

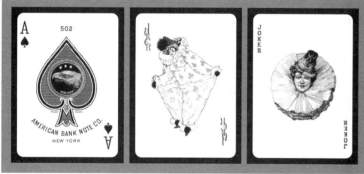

RU14

RU14 AMERICAN BANK NOTE #502
New York, c1908. The company, a long established firm, known for its fine printing of currency, stock certificates, and other beautifully detailed work, manufactured playing cards for a brief period. They only used one Ace of Spades design, sometimes numbered or named as in this issue, but more often just plain. The first Joker was used by ABN in most of their decks and was later adopted by Russell. The second is much rarer. Other brands include Steamboats #105, Bank Note #91 and Liberty #201. All regular 2 1/2 x 3 1/2 inches ABN cards will be given this number.

RU17

RU17 RUSSELL PLAYING CARD CO.
New York, c1912. The Russell Co. also used only two Aces for their entire existence. Most were plain, but some had the brand name written across the top. The first Ace was like the Kalamazoo Ace RU13, but with the 'K' changed to an 'R'. The few decks we have seen with this Ace have a similar Joker to RU3.

RU18 Ace

**RU18 RUSSELL PLAYING
CARD CO.**
*c1912. This was their most
popular Ace. Many of the same
brands that sold well under
Kalamazoo were still packaged in a
Kalamazoo box, but came with the
new Russell Ace. The Steamboat
Joker is from a brand called
Steamboat 7-11 by Kalamazoo PC,
but with this Russell Ace. Many of
the brands that came with this Ace
were wide, some were narrow and
some came in both sizes. The wide
ones particularly, came with
special Jokers and a number of
these are shown here.*

RU18 Jokers

There were many brand names used by Kalamazoo and Russell through the years, which never had their own Aces. They were identified by the Jokers and/or the box. The Aces used were RU12, RU13, RU17 and RU18. Russell Playing Card Co., especially, even after it became a division of USPC in 1929, produced a large number of brands using the RU17 or RU18 Spade Aces. These included:

Aristocrat #727	Aristocrat Whist #414	Aristocrat Pinochle #626
Autocrat #828	Battle Axe #822	Beauty
Beauty Pinochle	Blue Ribbon Pinochle #929	Broadway #288
Bridge Whist #60	Bridgit #227	Chancellor Pinochle #64
Conqueror #388	Cosmos	Cricket #03
Double Skat #632	Emperor Pinochle #244	Enterprise #14
Eureka #010	Fads & Fancies #150 and #525	Fifty-Fifty
Germania Pinochle	Hawk #0000	Idle Hour #811
Knickerbocker Narrow #227	Lustre (Metro PCC, NY)	Manhattan Pinochle
Mogul #100	Mogul Pinochle	New Deal
Pastime #712	Recruits #76	Regents Narrow #390
Rustlers	Rustlers #136	Rustlers Pinochle #118
Smart Set	Square Deal #06	Square Deal Pinochle
Success Pinochle #148	Whoopee #128	

And probably many more! All of these decks are grouped as RU17 or RU18.

An interesting observation - Russell always listed their address as New York, but they never made a card in any location other than Milltown, New Jersey. However, after the purchase by Rosenthal, they maintained offices in New York City, first at 346 Broadway and then at 200 5th Avenue.

RU19

RU19 ARISTOCRAT
Russell, New York, from 1915 to the present. This deck was adopted from the American Bank Note styling. It adopted the Joker, the backs and the quality from RU16 and was chosen by discriminating card players for years.

RU21a

RU21a LILY STEAMBOAT #66
Kalamazoo Paper Box, c1904. A similar Lily Ace of Spades came with a Steamboat Joker (like RU9) and box.

RU20

RU20 STEAMBOAT #7-11
Kalamazoo Playing Card Co., New York, c1911. A similar deck to RU13 with its Steamboat Joker but in a Kalamazoo PCC box with a New York address! Before introducing the Russell Ace to the old RPC brands, the new company must have issued this 'no-name' Ace of Spades. This Ace was also used with a RU18 Joker without the 'R' in the corners. Finally, a narrow version named Whist #811 comes with a somewhat different generic Ace of Spades and the common Russell Joker.

Other decks from Kalamazoo Paper Box and Card Co. and Russell Card Co. which have come to light subsequent to the publication of Volume IV in 1980 are listed as follows:

RU22

RU22 CRICKET
Kalamazoo Paper Box, c1904. Cricket was produced with at least four different Aces and two Jokers. In addition to this Ace and the one shown below (RU22a), Cricket was also made with the standard KPB Ace (RU12) and Cricket Joker and the Russell (RU18) Ace with the Cricket Player Joker.

RU22a

RU22a CRICKET
Kalamazoo Paper Box, c1904. A Lily quality deck with the Cricket Joker (RU12).

RU21

RU21 LILY
Kalamazoo Paper Box c1904. The Lily quality decks also had their own brand, in addition to being used on all other KPB brands of the time. Note the lily theme on the Ace, back and Joker.

RU23

RU23 CHANCELLOR CLUB #228
Kalamazoo Paper Box, c1904. One of KPB's more luxurious brands came with the Chancellor Joker like RU12.

RU23a

RU23a CHANCELLOR CLUB #228
Kalamazoo Paper Box, c1905. A different version with the portrait replaced by a monogram. The same Joker was used with both decks. The name 'Chancellor' was also used later by Russell on a Pinochle deck #164.

RU24

RU24 PINOCHLE #966
Kalamazoo Paper Box, c1904. The interior of the Spade pip on the Ace of Spades has the 'success' shield containing the motto 'Successus Success' (refer to the RU11 Joker). The 'Whig King' Joker shown was used with a number of Kalamazoo/Russell decks.

RU25

RU25 BISMARCK PINOCHLE #148
By KPB. This 48 card deck came in a luxurious slip case with a photo of Bismarck on the case. The same picture was used for the Chancellor Joker on RU23 and RU23a.

RU26

RU26 LILAC WHIST #333
Kalamazoo Paper Box, c1906. Despite the early period of this company, they manufactured a narrow whist size deck. The tax stamp is dated June 1, 1906.

RU26a LILAC WHIST #151 *(not pictured)*
Kalamazoo Playing Card Co., c1910. The KPC version of RU26.

RU26b LILAC BRIDGE WHIST *(not pictured)*
Russell Playing Card Co., c1913. A slightly later version, this time with the RU18 Ace and the RU26 Joker changed to 'Bridge Whist'.

RU27

RU27 FADS AND FANCIES #150
Kalamazoo Playing Card Co., c1910. A whist size deck which came with both standard KPC Aces and a special Joker.

RU28 SQUARE DEAL #06
Kalamazoo Playing Card Co., c1908. A wide deck with the RU12 Ace and a special Joker.

RU28

RU29

RU29 MONTE CARLO #528
Kalamazoo Playing Card Co., c1907. A KPC deck with a named Ace of Spades reminiscent of RU12. A beautiful gold edged deck in a slipcase with a Joker featuring the Monte Carlo casino and a colorful picture back.

RU32

RU32 GOLDFIELD
Goldfield Playing Card Co. (KPB), c1905. Another deck by KPB with a name designed to sell the cards in a specific market.

RU30

RU30 MAGYAR-HELVET-KARTYA #68
Kalamazoo Playing Card Co., 1911. This Hungarian Seasons pattern deck has no identity markings on the face of the cards. Without its wrapper it would have joined the ranks of anonymous European decks.

RU33

RU33 STEAMBOAT #209
St. Louis Playing Card Co. (KPB), c1905. KPB made at least three different brands under the St. Louis Playing Card Co. name – likely for sale in that area of the country.

Kalamazoo Paper Box and Card Co. and successors KPC and Russell, also made playing cards under a number of other names. Further research since publication of the Encyclopedia in 1980 has unearthed the following companies. Undoubtedly more will surface in the future.

RU31

RU31 UNSERFRITZ PINOCHLE CARDS #700
Manhattan Playing Card Co. (KPB), c1905. This company also issued at least two other decks packaged in Manhattan Playing Card Co. boxes, Victor and Manhattan Pinochle, both with the RU18 Ace.

RU34

RU35

RU34 CHESTER #77
St. Louis Playing Card Co. (KPB), c1905.

RU35 REGAL #55
St. Louis Playing Card Co. (KPB), c1905. The three St. Louis Playing Card Co. decks all used the pictured Joker.

RU36

RU36 FORTUNA
Columbia Playing Card Co. (KPB), c1905. Another name used by KPB in the St. Louis area.

Universal Playing Card Company is also a bit of a mystery and we found pictures of the following decks amongst Gene's files. In the original Volume VI, Universal had been identified as a Kalamazoo company and Gene had indicated that the Joker on the next listing was the same as RU18. In addition, the Joker on RU37b is the same as RU24 demonstrating that this was a company that was either formed as part of the Russell group or later acquired by them.

RU37 STEAMBOAT #9999 *(not pictured)*
Universal Playing Card Co. (KPC), New York, c1911. This deck had an anonymous Ace of Spades, but the KPC/Russell RU18 Steamboat Joker. Another Steamboat #9999 was made by Ferd. Messmer Mfg. Co., 10 South Broadway, St. Louis in 1906. It had a KPB tax stamp on the box.

RU37a PASTIME #712 *(not pictured)*
Universal Playing Card Co. (KPC), New York, c1910. An advertising deck for A.D.S. which came with the Joker below.

RU37b

RU37b ST. REGIS #387
Universal Playing Card Co., New York, c1905. This deck also has A.D.S. on the Ace of Spades and the Whig King Joker of RU24.

RU37c

RU37c UNIVERSAL PLAYING CARD COMPANY
New York, c1910. This Ace is quite different and it comes with a previously unknown Joker.

RU37d

RU37d KEY
Universal Playing Card Company, New York, c1910. Another Universal Ace but, unfortunately, no Joker was found.

RU38

RU38 BROADWAY #288
(Formerly MSW101), Metropolitan Playing Card Co., New York, c1905. This deck was first produced by KPB and subsequently by Russell. The Joker is identical to that of St. Louis Playing Card Co., RU33.

RU38a

RU38a PRINCE HENRY PINOCHLE #123
(Formerly MSW102), Metropolitan Playing Card Co., actually KPB, c1905.

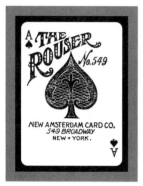

RU38b THE ROUSER #549
*New Amsterdam Card Co., New York,
c1905. Another name used by
KPB/KPC for local market sales. This
deck also uses the Jester Joker of St.
Louis Playing Card Co.*

RU38b

It should be noted that The Bay State Card Company had a connection with Russell that was demonstrated by Gene Hochman in Volume VI of the original Encyclopedia. As we believe that Bay State originally was part of Standard PCC, its decks are listed in Chapter 12.

Undoubtedly more decks made by otherwise obscure companies will be found and traced to Kalamazoo Paper Box and its successor companies. Similarly, other decks by KPB/KPC/Russell will turn up and be documented in the future.

To complete the story of Benjamin Rosenthal and the Russell Playing Card Co. we return to the Russell/Kalamazoo merger. In the years following this merger, Mr. Rosenthal bought out all of his original associates and acquired nearly all of the outstanding stock. As time progressed, many offers were presented by USPC to merge the Russell business into their conglomerate, at that time the largest producer of playing cards in the world. Finally it was decided late in 1929 that Russell Playing Card Co. was to enter the USPC fold. Mr. Rosenthal became the largest individual stockholder in the giant company and was appointed Executive Vice-President, Chairman of its Executive Committee, and a member of the Board of Directors. He also retained the Presidency of Russell, which was to continue to operate as a separate entity.

In 1936, The United States Playing Card Co., announced through Mr. Benjamin Rosenthal, Executive Vice-President, that the Milltown plant would be closed and the company moved to the main factory and offices of USPC in Cincinnati. Certain Russell brands would continue to be manufactured and, indeed, Russell Playing Cards, principally Blue Ribbon and Aristocrat, are still being made by USPC today.

STANDARD PLAYING CARD MANUFACTURING COMPANY

Standard Playing Card Co. (SPCC) commenced the manufacture of playing cards in Chicago about 1890. Shortly thereafter they were purchased by the United States Printing Co. and became one of the companies that formed the original United States Playing Card Co. in 1894. Standard Playing Card Co. continued to operate independently in Chicago and remained active until it was merged into the Consolidated-Dougherty Corp. in 1930. SPCC always had Chicago as the location on their Aces of Spades, and in latter years the company's address was known to be 412 Orleans Street in that city.

It is interesting to note that during their entire existence, SPCC never showed a brand name on any one of their Aces of Spades. There were three known Aces used for their wide issues, all of which are shown here. Initially it was thought that there was only the one wide Joker ('I've Got Him' pictured below) but several others, identified with specific brands, have surfaced and are shown in this chapter.

SU1

SU2

SU1 STEAMBOATS #900
Standard Playing Card Co., Chicago, c1890. This Ace of Spades, the first, came with all their brands in the early 1890s. The one pictured is from a Steamboats deck that featured an early Steamboat Joker. The other very rare Joker known to come with this Ace is also pictured.

SU2 STANDARD PLAYING CARD CO.
Standard Playing Card Co., Chicago, c1895. A slightly later Ace of Spades was used for Airship, Steamboats, and other decks.

SU3   SU4

SU3 STEAMBOATS #900
Standard Playing Card Co., Chicago, c1900. This Ace was made with both a color and black and white monogram. All decks with the black and white monogram are classified as SU3.

SU4 SOCIETY #1000
Standard Playing Card Co., Chicago, c1900. The color version was soon in use and is by far the most common of the SPCC Aces of Spades found today.

Many decks using the above Aces came with special Jokers and we have pictured ones here for Society #1000, Roundup #902, Liberty Bell #1776, Peerless #304 and Lucky Draw #905. The Peerless Joker has the same picture as its multi-colored back, reminiscent of certain of the Congress #606 wide named decks.

EARLY STANDARD PCC JOKERS

Over the years SPCC used a large number of brand names, but they appeared only on boxes and wrappers. These brands used different Aces at different times, depending on date of manufacture. A listing of the SPCC brands, taken from catalogs over the 1906-1927 period, included:

American Beauty #3333	Auction Bridge Whist #104	Aviator #914
Bank Note #91	Banner #399	Bird #809
Bismarck Pinochle #904	Bon Ton #906	Celutone
Colonial	Corona #377	Eagle #940
Earl #10	Good Luck #910	Kaiser Wilhelm Pinochle #21
Kenilworth	Kodak	Liberty Bell #1776
Lucky Draw #893	Monitor Steamboats #900	Oxford #59
Paramount #1003	Peerless #304	Ping Pong #912
Pinochle #9061/2	President #106 (Pinochle #107)	Radio #648
Radium #612	Renown #901	Rhineland Pinochle
Riviera	Roundup #902	Rugby
Saratoga #11	Society #1001	Society (Plain Edge) #1000
Society Pinochle	Steamboats #898	Special Steamboats #899
Stand Pat #25	Suffragette #4000	Teddy Bear
Uncle Sam #950		

SPCC also had a number of wide, named backs that were issued with their Society brand. These were of the monotone style concurrently used by NYCC and Kalamazoo Playing Card Co. The one pictured under SU4 is entitled 'It Listens Good'.

SU5

SU5a

SU5 AIRSHIP #909
Standard Playing Card Co., Chicago, c1900. Separate listings have been made for two special issues that featured unusual back designs. A most desirable issue, in great demand, is Airship #909. It was normally made with the SU2 Ace although the deck was also made with the SU3 Ace. It usually featured the 'I've Got Him' Joker, although the special airship Joker pictured here came with some early issues.

SU5a FAST MAIL #44
Standard Playing Card Co., Chicago, c1900. A similar back, but with a locomotive instead of an airship. This deck also appeals to the many railroad collectors. It came with both the SU2 and SU3 Ace of Spades and the standard Joker.

SU6 NATIVE LAND SERIES CO. #2000
Native Land Series Co., Chicago and San Francisco, c1900. Only one deck in this 'series' has been located although it is possible there were other backs, perhaps also with a patriotic theme. The cards were clearly made by SPCC.

SU6

Standard also made a limited number of standard Aces of Spades for their narrow whist size cards.

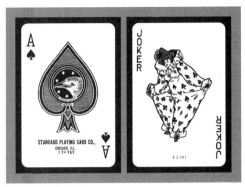

SU7

SU7 AMERICAN BEAUTY #3333
Standard Playing Card Co., c1910. This was the popular Ace of Spades during the Auction bridge era. Note the startling resemblance of the Ace of Spades and Joker to American Banknote Co.'s narrow whist deck RU16. In fact, an advertisement of that period entitled 'The Plain Truth' exhorts customers of the American Bank Note Co. not to buy the "cheap and inferior imitation Bank Note #91" (by SPCC) but rather their genuine brand #251.

SU7a

SU7a AMERICAN BEAUTY #3333
Standard Playing Card Co., c1905. A wide version of this brand has recently come to light.

SU8

SU8 AUCTION BRIDGE
Standard Playing Card Co., c1910. The second Ace of Spades, perhaps as a result of pressure from ABN, shows a far more graceful bird and a new Joker. This Ace was made in color as well as black and white.

SU9

SU9 STANDARD PLAYING CARD CO.
c1930. Another very plain, narrow Ace, with the company name printed under the spade pip, plus a very colorful Joker.

SU9a

SU9a AVIATOR
Standard Playing Card Co., c1930. While this deck has a generic Ace, it came wrapped with a SPCC tax stamp.

SU9b

SU9b STANDARD PLAYING CARD CO.
c1930. This deck has another plain Ace with no definite identification. Based on its style we believe it likely belongs in SPCC.

SU9c

SU9c STANDARD PLAYING CARD CO.
c1923. This narrow deck, with a plain Ace, came in a
Standard box entitled Auction Bridge Whist and has
company identification on the Joker.

SU12

SU12 LIDO
Standard Playing Card Co., c1925. A colorful Ace of Spades for whist
size decks with named art deco backs.

SU10

SU10 BUSTER #10
Standard Playing Card Co., c1900. The company
made two miniature decks, both with scaled down
versions of the SU2 Ace of Spades and the 'I've got
him' Joker. This one was the larger of the two,
measuring 1 11/16 x 2 7/16 inches. It was clearly
designed to compete with USPC's Fauntleroy cards.

SU13

SU13 AUSTEN BEAUTIES #506
(Formerly MSW33), J.I. Austen Company, Chicago, c1895. Another
manufacturer, subsequently identified as SPCC, is now listed in this chapter.
This company was the publisher or distributor of several decks, including this
Austen Beauty.

SU11

SU11 JAP #20
Standard Playing Card Co., c1910. A smaller patience
deck based on the same cards as SU10. The size of this
deck was 1 1/4 x 1 3/4 inches, the same as USPC's
Little Duke.

SU14

SU14 J.I. AUSTEN COMPANY
(Formerly MSW34), Chicago, c1895. Another standard deck made for
this publisher.

SU15

SU15 STEAMBOATS #199 1/2
(Formerly MSW72), Crescent Playing Card Co., Chicago, c1900. Note the resemblance to National's Crescents Ace of Spades (NU11). The Joker is in color.

SU17 *SU17a*

SU17 ALBERT PICK & COMPANY
Chicago, c1905. Another name that has proven to be one used by SPCC. The Joker is the same as the one used by Crescent PCC (SU15).

SU17a FASTMAIL
Albert Pick & Company, c1905. A different Ace of Spades came with a 'Fastmail' back and the SU15 Joker.

It is now known that Crescent Playing Card Co. was a name used by Standard (after its take-over by USPC). Comparisons of the court cards and the fact that a Standard PCC Banner #399 was recently found along with an identical (except for the Ace of Spades) Crescent PCC Banner #399 deck have provided some of the evidence needed.

Only one Crescent deck was depicted in the earlier Encyclopedia and it has been moved to this section. Crescent PCC also made the following brands (probably with the SU15 Ace) according to an old price list, c1905, when they were located at 88-96 East Ohio Street, Chicago: Banner #399, Electric #666, King William Pinochle #499, Monarch #299, Rhineland #555, Steamboat #00, #199 and #199 1/2, Trolley #599 and Whist #699.

SU18

SU18 BELLEVUE
(Formerly MSW106), Parker Brothers, Salem, Mass., c1900. The largest manufacturer of games entered the playing card field about 1900. It has now been determined that the manufacturer of Parker Bros. cards was Standard Playing Card Co. of Chicago. These decks came with named monotone backs, no doubt intended to compete with Congress #606.

SU16

SU16 CRESCENT PCC
Chicago, c1900. A different Ace Spades and Joker were used in this rare deck.

SU19

SU19 REMBRANDTS
(Formerly MSW107), Parker Brothers, Salem, Mass., c1900. Another brand made for Parker Bros. by SPCC. These decks also had named backs, this time with matching named Jokers like certain Congress styles.

In Chapter 4, Longley, we mentioned that The Bay State Card Company had a connection with Card Fabrique, Globe, North American Card Co., etc. as the earliest Bay State Ace of Spades was clearly derived from L22. In addition, the courts of that deck were the same as those of L27. Whether one of those companies was sold to someone who started Bay State, or whether Card Fabrique/Globe started it is unclear. What we do know is that by comparing courts, we can be sure that Bay State became part of Standard Playing Card Company of Chicago in the early 1890s and thereby part of the USPC fold in 1894.

Bay State Playing Card Co. and Buckeye Playing Card Co. were identified as the same company in Volume IV of the Encyclopedia and originally listed as MSW36 to MSW46. We have seen Bay State decks with Standard PCC, NYCC, USPC and Russell PCC tax stamps (the connection with Russell was demonstrated by Gene Hochman in Volume VI of the original work – see SU26) and debated which chapter would be most appropriate for its inclusion. Because it was clearly associated with Standard in its early days, we decided to reclassify the cards of Bay State and Buckeye with SU numbers starting at SU21.

SU23a

SU23a WHIST CLUB #37
The Bay State Card Co., c1895. A narrow version of this Ace of Spades with the same stylish courts used by the North American Card Co.

SU23b

SU23b THISTLE
(Formerly MSW46), Bay State Card Co., c1905. A narrow version with a plainer Ace, manufactured as a Pinochle deck.

SU21 BUCKEYE CARD CO. *(not pictured)*
(Formerly MSW36), c1900. The standard Ace of Spades, probably, like Bay State, used as the Ace for several different brands. It is identical, except for the name, to SU24.

SU22 BUCKEYE CARD CO. *(not pictured)*
(Formerly MSW42), c1900. A different Ace of Spades and Joker used by this company. The identical Ace of Spades, Joker, etc., to SU28 was used except 'Buckeye' replaces 'Bay State'.

SU24

SU24 BAY STATE CARD CO.
(formerly MSW37), c1900. This is the Bay State equivalent to SU21 and the Ace of Spades shown here was from their brand Steamboat #09. There was no Joker found with this pack, but the tax stamp read 1907. Most Bay State decks have brand names only on the boxes, so identification is impossible without the box. Another interesting brand to use this Ace was Four-in-Hand #22 with the box depicting a carriage carrying a smartly dressed couple being pulled by four prancing horses.

SU23

SU23 THISTLE
(Formerly MSW45), Bay State Card Co., c1895. The Ace of Spades and the courts have a marked similarity to those of Card Fabrique/Globe and it is likely their earliest Ace. This Ace of Spades came with both regular courts and the special courts used by North American Card Co. (L27). It was also used for a brand named 'Colonial' which had its own branded Joker.

SU25

SU28

SU25 NUGGETT J
(Formerly MSW38), Bay State Card Co., c1900. This brand was made with the SU24 Ace as well as this interesting anonymous Ace of Spades. It had a special 'Prospector' Joker.

SU28 BAY STATE CARD CO.
(formerly MSW41), c1900. A different Ace of Spades found without a box so remaining unnamed. This is the Bay State version of the second Buckeye listing (SU22). It was also produced as 'Judge 97' and the Joker from this deck, which is similar to RU33, is also pictured.

SU26

SU28a

SU26 JUDGE
(Formerly MSW39), Bay State Card Co., c1900. This brand used a different Ace of Spades but with the same cupid motif. It was this Ace, compared with the identical Butler Brothers Ace, also pictured, but by Russell Playing Card Co., which was the final piece of evidence necessary to establish that Russell and Bay State/Buckeye were all in the same family. It should also be noted that this deck with the identical Ace of Spades and Joker has been found in an Eclipse #432 box made by NYCC of Long Island City.

SU28a STEAMBOATS
Bay State Card Co., c1905. A recently discovered Steamboats version of the SU28 Ace with a Steamboat Joker identical to SU1.

SU27

SU29

SU27 MAGICIAN BRAND J
(Formerly MSW40), Bay State Card Co., c1900. The SU24 Ace of Spades was used but the deck had a special Magician Joker.

SU29 REGAL
(Formerly MSW43), Bay State Card Co., c1905. This is perhaps the only Bay State deck with the brand name on the Ace of Spades. The courts are similar to those of NYCC.

SU30

SU31

SU31 STAMPEDE #189
Bay State Card Co., c1910. This deck has an unnamed Ace of Spades but the box has the maker's name. The tax stamp is dated 1911 and has a USPC cancellation.

SU30 VALOR
(Formerly MSW44), Bay State Card Co., c1910. A narrow brand made by this company.

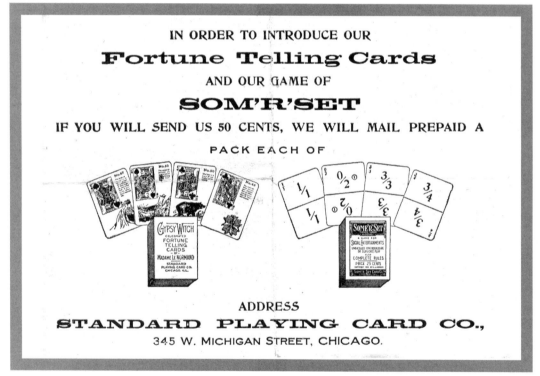

SPCC ADVERTISEMENT

OTHER MAKERS OF WIDE CARDS

A large number of the decks previously listed in Part 4 in the Other Makers section (MSW), have found a home in one of the other Chapters. To assist those who are familiar with the original Encyclopedia, we prepared the following schedule that gives the original listing and its new reference.

The reader can see, for example, that MSW1 to MSW18 are now in Chapter 4, Longley Brothers and Successor Companies. In addition, a number of miscellaneous, but we believe related, manufacturers were also moved to this chapter due to their relationship to the Longley group. Finally, we determined that several other decks in the MSW section were made by Russell, USPC, Dougherty, Standard or NYCC, so we transferred them to the appropriate chapter.

For ease of reference, where we have moved a listing to another chapter and changed its listing number, we have provided a reference to its old number (e.g. "formerly MSW11"). Finally, we have added a few new listings to catalogue new discoveries.

NAME	ORIGINAL ENCYCLOPEDIA	NEW ENCYCLOPEDIA
American PCC	MSW1 to MSW18	L52 to L69
American PCC- New York	MSW22, 23	L9, 10
Astor PCC	MSW27	Same
Atlantic PCC	MSW29 to MSW31	US49 to US51
J. I. Austen	MSW33, 34	SU13, 14
Buckeye/Bay State	MSW36 to MSW46	SU21 to SU31
Blackstone	MSW48, 49	L39, 39a
Brooklyn	MSW51	L38
Card Fabrique	MSW53 to MSW59	L14 to L18
Caterson & Brotz	MSW63 to MSW67	L32 to L36
Reynolds	MSW68	L37, 37a
Chicago Card Co.	MSW69	L24
Continental	MSW70	MSW79a
Crescent	MSW72	SU15
Comet	MSW74	AD50
T. L. DeLand	MSW75	Same
Dorrity	MSW78	Same
Excelsior	MSW80 to MSW82	L29 to L31
Empire State	MSW85	Same
Hornet	MSW87	US46
Independent	MSW88, 89	Same
Koehler	MSW91	Same
Knickerbocker	MSW93	L38a
Lighthouse	MSW95, 96	NY81, 81a
Canary	MSW97	NY78
New England PCC	MSW99	Same
Broadway	MSW101	RU38
Metropolitan	MSW102	RU38a
American Beauty	MSW103	L27
Owl & Spider	MSW105	US48
Parker Bros.	MSW106, 107	SU18, 19
Popular PCC	MSW110	NY77
Pyramid	MSW112 to MSW122a	PY1 to PY12
Union/Eureka	MSW123 to MSW130	L40 to L49
U. S. Card Co.	MSW132	MSW91a
Tower Mfg.	MSW133	US53
Waldorf	MSW134	AD47
Western Press	MSW135,136	Same
Western PCC	MSW137	L28c

MSW27

MSW27 ASTOR PLAYING CARD CO
*New York, c1920. There is also an
'Astor' in the narrow chapter (MSN24)
but they do not seem to have been made
by the same company.*

MSW78a

MSW78a STEAMBOAT #1999
*Dorrity Card Manufacturing Co., New
York, c1903. This deck has a small
plain pip on the Ace of Spades and a
great Joker advertising the company.*

MSW75

MSW75 THEODORE L. DELAND
*Philadelphia, 1914. Gene Hochman believed Deland published regular
playing cards during the early part of the 20th century, before devoting
his efforts to the famous trick and marked cards dealt with in depth in
Chapter 33 (see N2 to N6). The deck pictured here is the same as the
one originally pictured but our deck is his very early 'Dollar' deck.*

MSW79

MSW79 LUCKY DOG INDIAN BRAND
*Dorrity Card Manufacturing Co., New York, 1905. The Joker is
identical to that of MSW79a.*

MSW79a

**MSW79a LUCKY DOG INDIAN
BRAND**
*Continental Playing Card Co. (formerly
MSW70), New York, c1906. This deck,
the only standard one known by this
company, came in a Pinochle version of
48 cards as well a 52 + Joker brand. The
Joker pictured is identical to the Joker in
the last Dorrity deck (see MSW79) and
the spade pip portion of the Ace is also
the same. In fact, a recent discovery
confirms that Continental was a
successor to Dorrity and made cards in
1906. Other known brands are Funny
Spot (see T13) and Peacock #20.*

MSW78

MSW78 PINOCHLE #1909
*Dorrity Card Manufacturing Co., New York, c1903. Dorrity is now
known to have operated in New York, but only for a brief period from
some time in 1903 to 1905. Other brands not listed here included
Casino, Reliance #3909 and Senate. This Pinochle deck is from the
USPC collection in Cincinnati. Note that the King of Hearts does not
have the usual sword behind his head.*

MSW85

**MSW85 THE EMPIRE STATE
PLAYING CARD CO.**
*Rochester, N.Y., c1920. The only deck
known by this maker is missing the
Joker.*

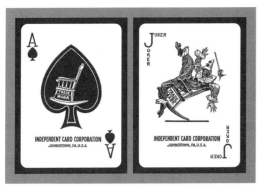

MSW88

MSW88 INDEPENDENT CARD CORPORATION
Johnstown, Pa., c1926. A well designed, unusual deck by this seldom heard of company which manufactured from 1926 to 1930. It features packaging in a metal case. Brands made by this company included Cambria #66, De Luxe, Freedom, Red, White & Blue, Rex #79, Tug-O-War Steamboat and Ricco Steamboat.

MSW91

MSW91 GOLD BACKGROUND PLAYING CARDS
By Jos. Koehler, Manufacturer, New York, c1885. This is a very unusual and beautiful deck, and the only example of cards we know of by this manufacturer is in the United States Playing Card Co. Museum. However, the following four decks appear to have been made by the same manufacturer as there are unmistakable similarities in either the Aces or the Jokers.

MSW88a

MSW88a TUG-O-WAR STEAMBOAT
Independent Card Corporation, Johnstown, Pa., c1926. The Ace on this deck does not give the name of the company.

MSW91a

MSW91a COLUMBIA #27
D. Lesser & Son, c1885. In this case the Joker, except for the circle with different initials, is the same as the Koehler Joker. To confuse matters more, a notation with this deck refers to Progress Playing Card Co., New York.

MSW89

MSW89 REX #79
Independent Card Corporation, Johnstown Pa., c1926. Another deck by this company. The cards are well made and it is surprising that this company did not leave a mark of more success. Other than the three decks listed, the only trace of this company we have found is a share certificate dated 1926.

MSW91b

MSW91b D. LESSER MANUFACTURER
c1885. A very different Lesser Ace, which is similar to the Koehler Ace, with a very unusual Joker.

MSW91c

MSW91d

MSW91c YALE #49
Liberty Card Co., New York, c1890. In this case the Ace of Spades has clear similarities to both the Koehler and second Lesser Aces.

MSW91d UNITED STATES CARD CO.
(Formerly MSW132), New York, c1880. The similarities between this Ace and that of the Koehler deck, together with the same Joker as MSW91b, demonstrate that the same company made them.

MSW99

MSW99 NEW ENGLAND PLAYING CARD CO.
Boston, Mass., c1925. Another mystery, probably made by an, as yet, unidentified larger company.

There has been much speculation about the origins of Western Playing Card Co., including whether American Playing Card Company of Kalamazoo, Michigan was one of its ancestors. While questions still remain, some pieces of the puzzle have come together.

The following table details the names, dates and addresses that we can be sure about.

Western Press	Chicago	1918
Logan Printing House	Chicago	1919
Midland Playing Card Co.	3652-66 Milwaukee Ave., Chicago	1920
	Lafayette, Ind.	1921
	1733 Irving Park Blvd., Chicago	1922
Inter-Ocean PCC	Same as Midland	1920-24
M. B. Sheffer Card Co.	3670-72 Milwaukee Ave., Chicago	1924
	Racine, Wis.	1926
Western PCC	Racine, Wis.	1927 on

Decks by Western Press and Logan Printing House are scarce, with ones from Midland, Inter-Ocean and Max B. Sheffer somewhat more common. Many examples of decks from Western PCC made from 1927 on can still be found.

In the period from 1918 to 1927 the companies made the following brands, and perhaps others.

Betsy Ross	Universal	Olympic
Western Steamboat	Golf	Matchless
Elite	Midland Special	Molly O
Ritz Carrollton	Dolly Madison	Martha Washington (Whist)
L. G. Special (Whist)	Invincible	Crusader
Peerless	Excello	Cosmos
Marvel Steamboat	Dauntless	What-Cheer

Some of these brands, Matchless and Elite for example, were made without identification of the manufacturer on any of the cards.

Note that one of the brands was Golf, a very popular brand over the years for American PCC. It is also noteworthy that some of the above brands, at least in their early versions, had the same courts as APPC, and that the later cards by Western PCC clearly copied the American Ace of Spades (see MSW135). What was the connection? Perhaps Max B. Sheffer, the first President, bought the business from American after WWI, or perhaps he purchased their equipment and plates.

The group continued in business until 1927, when the Western Playing Card Company was formed on August 15th. Their first price list contained the same brands previously listed by Sheffer, indicating it was probably just a change of name, not a take-over by another company. The following listings include more designs that have turned up since the original publication. No doubt there are many more that will be discovered in years ahead.

MSW131b

MSW131b GOLF
Midland Playing Card Co., Chicago, 1920. This deck has a generic Ace and the Joker, although only in yellow and black, that was to become the standard for the Western PCC decks (MSW136).

MSW131

MSW131 MIDLAND SPECIAL
Max B. Sheffer/Midland Playing Card Co., Chicago, 1920. The Ace of Spades says Max. B. Sheffer and the box names the deck and says Midland Playing Cards. The courts are identical to those of American PCC.

MSW131a GOOD TIMES
Inter-Ocean Playing Card Co., Lafayette, Indiana, c1920. A pinochle deck made by another of the Sheffer/Midland companies identified only on the box.

MSW131a

MSW135

MSW135 THE WESTERN PRESS
Chicago, c1918. This Ace of Spades has the name "The Western Press", Chicago and a similar parrot Joker to that of MSW131. Two other Jokers known to come with this Ace of Spades are pictured.

MSW136

MSW136 UNIVERSAL PLAYING CARDS
Western Playing Card Co., Chicago, c1927. A later deck from the successor company to Sheffer with a plain Ace of Spades. This deck comes with a black and white Ace of Spades and a tax stamp cancelled WPL CO. The Joker is red, blue and yellow and the box only gives the name of the deck, with no mention of the company. A version of this deck on cheaper stock comes in a Marvel Steamboat #123 box and this deck is also known to have come with other brand names in a Western PCC box.

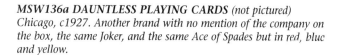

MSW136a DAUNTLESS PLAYING CARDS *(not pictured)*
Chicago, c1927. Another brand with no mention of the company on the box, the same Joker, and the same Ace of Spades but in red, blue and yellow.

A number of other decks have been 'discovered' since publication of the original Encyclopedia and, where they do not seem to fit with makers already listed, we have given them new numbers starting with MSW140 to avoid confusion with the original listings.

MSW140 BUNKER HILL #1776
Oxford Playing Card Co., Boston, c1910. The only deck we have seen from this company has an advertising back but the style of the Ace and the brand number suggest they made standard cards as well.

MSW140

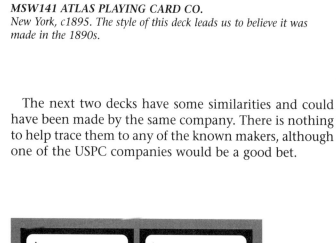

MSW141

MSW141 ATLAS PLAYING CARD CO.
New York, c1895. The style of this deck leads us to believe it was made in the 1890s.

The next two decks have some similarities and could have been made by the same company. There is nothing to help trace them to any of the known makers, although one of the USPC companies would be a good bet.

MSW142

MSW142 OVERLAND
c1900.

MSW143

MSW143 ROUGH RIDER
c1900. A deck pertaining to the activities of Teddy Roosevelt.

NARROW CARDS

Most of the playing card makers covered in the preceding chapters also made narrow (bridge/whist size) decks which, if identified with the maker, are listed and described therein. The decks listed in this chapter are mainly limited to those that originated during the 'bridge era', the early 1920s to the late 1930s, or to those made by the large manufacturers for another publisher.

We have included all the decks listed in the original Encyclopedia (although certain listings have been moved to other chapters) and others that have been brought to our attention. There were literally hundreds of publishers who placed their name and trademarks on the Aces of Spades of narrow decks during this period. Most were actually manufactured by the large makers

and when the actual maker is known the deck will be so identified. A few were made by very successful smaller manufacturers and some by relatively obscure companies that probably made only one or two brands during their relatively short life spans.

Gene Hochman believed it was not necessary to mention every brand, made by every maker, as was done with the earlier cards because, with a few exceptions, the value of these decks is relatively minimal. We still subscribe to this view. While we have added to his original listings, there is still much to do to document all narrow brands, especially if the period after World War II is included. We leave it to another researcher to complete this portion of the history of United States playing cards.

MSN1

MSN1 ARROW PLAYING CARD CO.
Chicago, c1930. This company was known for the Discus round cards (see 018). They had several different street addresses but remained in Chicago. Due to legal problems involving the 'Arrow' name, the company name was soon changed to Arrco.

MSN1b/MSN1c

MSN1b/MSN1c ARRCO PLAYING CARD CO.
Chicago. Numerous examples of this Ace of Spades, including these two with slight differences, have been found.

MSN1a

MSN1a ARRCO PLAYING CARD CO.
(formerly The Arrow Playing Card Co.), c1935. A transition deck made during the time of the name change.

MSN1d

MSN1d ARROW PLAYING CARD CO.
c1925. The clues to this unnamed deck are a tax stamp with APC on it and the close resemblance to the Arrco Ace of Spades.

MSN2

MSN2 SERVICE
Arrco, c1943. A wide deck produced for sale in USO canteens during WWII. It also came in a narrow version.

MSN4a

MSN4a ENARDOE
By Edward Drane & Co., Chicago, c1932. A standard deck, almost surely made by Arrco, but with the suit colors reversed, i.e., red clubs and spades and black hearts and diamonds. Apparently the idea was to slip one of these into your regular bridge game and confuse everyone. No wonder there are not many of these decks around. Incidentally, 'Enardoe' looks suspiciously like E. O. Drane backwards.

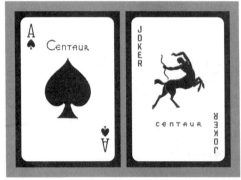

MSN3

MSN3 CENTAUR
Arrco, c1940. One of many brands manufactured by Arrco. The name of the company was only shown on the box.

MSN5/MSN5a

MSN5/MSN5a ARRCO
Chicago. Another relatively early Arrco design which came with either a named or generic Ace of Spades.

MSN4

MSN4 ARRCO
c1940 to 1970s. This Ace of Spades was issued in many different decks, including novelty, advertising as well as standard.

MSN6/MSN6a

MSN6/MSN6a ARRCO
Chicago. More Arrco designs.

MSN7

MSN7 ARRCO
Chicago, c1947. Another, relatively early, design.

Undoubtedly, there were a number of other Arrco designs which we have missed or that will be discovered in the future.

Bailey, Banks, & Biddle, a large Philadelphia jeweller and stationer had many different decks made to order by large manufacturers. Previously listed in this section, with the exception of MSN11 below, they have been moved to other chapters – refer to L75, NY82, NY82a and NY82b.

MSN11

MSN11 BAILEY, BANKS, & BIDDLE
c1935. This deck has the words "Play-well, No-glare" on the Ace of Spades.

The Brown & Bigelow Co. of St. Paul, Minnesota was one of the leading producers of playing cards in the United States from the late 1920s until the 1980s. They manufactured playing cards under many different names. In addition to bridge size cards, the company produced many unusual novelty decks and teaching decks, as well as a wide variety of advertising decks. Their ad decks normally had advertising on the boxes and the backs only, although some are known with special aces and/or jokers. They also manufactured the Stancraft Playing Cards and became a division of Standard Packaging Corp. In the 1940s, they also produced cards for (or under the name of) Nasco Playing Card Co., Chicago.

MSN12

MSN12 BROWN & BIGELOW
c1940. This Ace was widely used by Brown & Bigelow.

MSN13

MSB13 REMEMBRANCE
Brown & Bigelow, c1935. This Ace proves the connection of these two names.

MSN14

MSB14 REMEMBRANCE
Brown & Bigelow, c1935. A generic Ace by this prolific company with a standard B&B Joker.

MSN15

MSN15 BROWN & BIGELOW
St. Paul, c1945 to 1980s. All standard decks with this Ace of Spades are listed under this reference number. This is their most familiar Ace and Joker.

MSN16

MSN16 KENT
Stancraft. All standard cards with this ace are relatively current.

MSN17

MSN17 FORTUNA QUALITY
Brown & Bigelow, c1930. An example of one of the many brands made by Brown & Bigelow.

MSN18

N18 NASCO
Brown & Bigelow, Chicago, c1945. As mentioned, this was a B&B company.

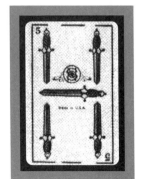

MSN19

MSN19 SPANISH
Nasco, Chicago, c1945. A Spanish suited deck by this maker.

MSN20

MSN20 NASCO
Chicago & New York, c1940.

MSN21

MSN21 NU VUE
Brown & Bigelow, 1960. New style courts and a special tint were the features offered in this new 'Modern Eye-Saving Concept'.

MSN22

MSN22 LUXOR
This is one of those mystery decks, which came with an early interesting Ace of Spades and Joker.

MSN23

MSN23 MONTE CARLO
The Buzza Company Inc., Minneapolis, c1930. Buzza were makers of a number of bridge accessories and their cards were made for them by USPC.

MSN24

MSN24 ASTOR
c1935. There is a strong resemblance to the Aces made by Cruver, so it is probable that this company also made this deck.

MSN28

MSN28 THE COLLEGIATE PLAYING CARD CO.
Washington, D.C., c1932. Another unusual brand almost surely made for use by colleges and universities.

MSN25

MSN25 EDWARD J. CADIGAN
New York c1940. As the Joker in this deck is identical to that of MSN1a, we can presume that Arrco made it.

MSN29

MSN29 CRITERION PLAYING CARD CO.
c1933. A standard pack made by the company that also made the No-Revoke deck (NR11).

MSN26

MSN26 CAIRO PLAYING CARD CO.
Chicago, c1935. Probably made by one of the Chicago manufacturers.

MSN30

MSN30 CRITERION PLAYING CARD CO
c1935. The maker of these Criterion decks has not been positively identified although it seems likely that they were made by Consolidated-Dougherty. This Ace of Spades came with and without the Criterion name.

MSN27

MSN27 CAROTTI PLAYING CARD CO.
10 E. 40th St., New York, home office, Sag Harbor, New York.

MSN31

MSN31 PERMANITE
By Cruver Playing Card Co., Chicago, c1935. An early plastic card which was marketed as washable.

MSN32   MSN33

MSN37

MSN32 CRUVER
c1935. This deck, with the words "Washable Plastic Playing Cards" written on the Ace of Spades further identifies this company's marketing strategy.

MSN33 ZODIAC
the DeBower Publishing Co., New York, 1931. Examination of the court cards reveals the maker of this deck to be Brown & Bigelow.

MSN37 KEM PLAYING CARDS
1935. These high priced, washable plastic cards were strongly promoted by Ely Culbertson, the noted bridge authority, and a part owner of the company. They proved to be very durable and many examples, in near perfect condition, are still found. One will find many minor variations in the aces and jokers over the years.

MSN34   MSN35

MSN34 DALE
Cruver Playing Card Co., New York. Due to the striking similarities between the Ace of Spades for this company and MSN31 and MSN32 there is little doubt that Dale was connected to Cruver.

MSN35 FAIRCO
E. E. Fairchild Co., Rochester, N.Y., c1930 to c1950. This company produced many varieties of bridge size cards featuring this Ace of Spades, albeit with minor variations. They also produced the Easibid teaching decks and many advertising packs.

The Gibson Playing Card Co., Cincinnati and New York produced cards starting around 1930. This smaller company produced a great many brands during the early contract bridge era. The Gibson PCC was part of the Gibson Greeting Card Co., Cincinnati. Eventually, one of the USPC executives left to join the Gibson Company. USPC subsequently bought the company, kept the executive and the playing card business and sold the greeting card business back to the original owners.

Note the similarities in all of the Aces of Spades (MSN39 to MSN50 and MSN65 to MSN67) produced by Gibson and shown below. In addition, the jokers for these decks always followed the motif suggested by the brand name (not all jokers have been pictured).

MSN36 FAIRCO
E. E. Fairchild Co., N.Y., c1940. Another style Ace of Spades from Fairchild.

MSN36

MSN39

MSN39 GIBSON SLIMS
Gibson PCC, Cincinnati. This deck was narrower (2 1/16 x 3 1/2 inches instead of the usual bridge size of 2 1/4 x 3 1/2 inches). It also came without the Gibson Slims name on the Ace of Spades in a standard Bridge size version.

MSN40

MSN40 FINESSE, *Gibson PCC, New York.*

MSN41

MSN41 CLASSIQUE, *Gibson PCC, New York.*

MSN42

MSN42 MINERVA, *Gibson PCC, New York.*

MSN43

MSN43 DIANA, *Gibson PCC, New York.*

MSN44

MSN44 HIGH STEPPER, *Gibson PCC, New York.*

MSN45

MSN45 SCOTTIE, *Gibson PCC, New York.*

MSN46

MSN46 CHALLENGE
*Gibson PCC, New York. The only wide
brand known to be made by Gibson.*

MSN46a

MSN46a CHALLENGE
*Gibson PCC, New York. This is the
narrow Ace of Spades with the same
name as the wide deck shown above.*

MSN47 THE FAMILY ALBUM
Gibson PCC, New York.

MSN47

MSN48

MSN48 BRILLIANTS, Gibson PCC, New York.

MSN49

MSN49 SILHOUETTE, Gibson PCC, New York.

MSN50 DICKIE BIRD
Gibson PCC, New York.

MSN50

The New York address on most of the Gibson brands confuses the issue. In addition, comparison with the Fan C Pack cards (King Press) shows a startling similarity in the design of the two companies' court cards. This suggests the possibility that King Press made some of the cards marketed by this company.

King Press, Hurley Playing Card Co., Fan C Pack and Play-Well were all the names used by this large manufacturer located in Carlstadt, New Jersey. They made cards during the contract bridge era in the 1930s and the 40s. Although the cards were made in New Jersey, their most popular brand, Fan C Pack, was listed as made in New York.

MSN51

MSN51 KING PRESS
Carlstadt, N.J., c1937. This Ace of Spades was used with several brands of cards produced by this company. The deck photographed was sold under the name 'Godey's Ladies Cards', and the backs had a reproduction of a fashion print from this popular Parisian fashion magazine.

MSN52

MSN52 PLAY-WELL
Hurley Playing Card Co., Carlstadt, N.J., c1935. Play-Well cards were patented and so the notation "Patents allowed and Pending" was shown on the Ace of Spades. To quote from an extra card packed in each deck, "No Glare (Easy on the Eyes). The face surface of each card is treated with many thousand microscopic dots which serve to absorb the destructive red rays of light so harmful to the eyes. This treatment of the card screens the otherwise intense white surface and reflects to the retina of the eye a softened tone that reduces eye fatigue and relieves nerve tension. The restful character of Play-Well cards will be apparent at once, and the absence of glare will be appreciated by all."

MSN53

MSN53 PLAY-WELL
*By King Press, Carlstadt, N.J., c1935.
The same deck as MSN52, but with the
other company name.*

MSN57

MSN57 GODEY'S LADIES CARDS
*Fan C Pack, c1932. This was the Fan C
Pack version of MSN51. The same backs
were used.*

MSN54

MSN54 PLAY-WELL
*A private brand made by the company for Young & Rudolph, a
stationer in Philadelphia. All private brands made by this company
can be included in this number.*

MSN58

MSN58 GOOFY PLAYING CARDS
*Fan C Pack, c1930. A strange deck with
the MSN55 Joker, perhaps with appeal to
those recovering from the effects of the
1929 crash.*

MSN55

MSN55 WEDGEWOOD
*Fan C Pack, New York, c1930. The
backs in all of the brands take their
styling from the name of the brand. The
Ace from this deck is like MSN62, with
the name Wedgewood on the Ace of
Spades.*

MSN59

MSN59 VICTORIAN LADY
Fan C Pack, c1930.

MSN60 GALAXY *(not pictured)*
Fan C Pack, c1935. The Ace was similar to MSN59.

MSN56

MSN56 LITTLE OLD NEW YORK
*Fan C Pack, 1925 to 1935. The backs of
the cards in this deck had scenes of New
York City in 1825. According to the box,
the theme was little old New York as it
was 100 years ago.*

MSN61

MSN61 MOSAIC
*Fan C Pack, c1935. The backs have a
mosaic tile pattern.*

MSN62

MSN62 PAST-L-EZE
Fan C Pack, c1935. One of the most valuable and most unusual of all of the standard, narrow, Bridge size decks. Some of these were made for them by E.E. Fairchild. The courts were newly designed and highly unusual in a light pastel shade of green, pink, or buff background.

MSN63 RAGGY SCOT *(not pictured)*
Fan C Pack, c1935. A similar Ace to MSN64 below.

MSN64

MSN64 POLKA DOT
Fan C Pack, c1936.

MSN65

MSN65 TECHNIQUE
Gibson Playing Card Co., c1925, New York City. Another deck manufactured by this company.

MSN66

MSN66 GOLDEN WEAVE
Although this Ace of Spades is without a company name it seems apparent by the design of this Ace that it is part of Gibson PPC.

MSN67

MSN67 PARTY
The same reasoning applies for this deck. Was Gibson using another name or did they manufacture this deck for a small, short-lived company?

MSN68

MSN68 KREKO
By La Crosse Playing Card Co., c1935. There are quite a number of these colored Aces of Spades in the same style as this one and the one following.

MSN69

MSN69 SCAMPER
By La Crosse Playing Card Co., c1937. Probably made by USPC.

MSN70

MSN70 THOROBRED PLAYING CARDS
By Stanley, Dayton, Ohio, c1940. These cards were probably manufactured by one of the major companies.

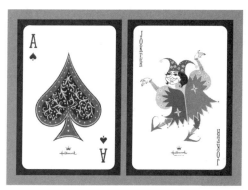

MSN71

MSN74

MSN71 HALLMARK
This deck, made by the same company that manufactures Hallmark greeting cards, was issued in 1940. Hallmark continued to make cards for a number of years.

MSN74 WESTERN PLAYING CARD CO.
Racine, Wis., c1933. A common narrow brand with one of their colorful jokers.

MSN72

MSN75

MSN72 PEAU-DOUX
A private brand made by USPC for Walgreen Drug Stores.

MSN75 CONTRABAND PLAYING CARDS
Chas. A. Stephens & Bros. Chicago, c1926. Another interesting deck, probably made by a large manufacturer for a stationer. The backs of this deck, issued during prohibition, show labels from a number of well-known liquor brands, thus the name. The great Art Deco Joker is in color.

MSN73

MSN76

MSN73 ITALIAN PLAYING CARDS
by Trinacria Playing Card Co., Chicago, 1910. This brand was made for Italian Americans living in this country. It was not intended for export.

MSN76 USPC
c1935. USPC made a number of decks for stores, etc., using this Ace of Spades, although usually it did not identify the company.

MSN77

MSN77 SPANISH ART
Sherman Brothers, Chicago, c1932. Again, likely made by USPC for this store.

MSN78

MSN78 J. E. CALDWELL & CO.
Philadelphia, USPC, 1917. Again made for a department store or stationer. Two Jokers we have seen with this Ace are shown.

MSN79

MSN 79 LAFRANCE
USPC, c1925. A deck made for 'The Fair', a Chicago store.

As mentioned, there is a great deal more work necessary to do full justice to the wide range of narrow decks made in America during the 20th century.

PICTORIAL BACKS

Many playing card collectors have a fascination with decks that have pictorial backs. These decks with a standard Ace of Spades and usually a standard Joker, normally have a colored back depicting a pretty scene, an attractive woman, an interesting building, etc. Sometimes the backs are of a single color, referred to as monotones, and often the backs are 'named', i.e. they have the name of the scene, person, etc. in inconspicuous letters at the bottom of the back of each card. From the late 1890s until the 1920s these were issued as wide decks. Narrow decks with similar backs were also made from the 1890s forward although their popularity was relatively low until the second and third decades of the 20th century.

While many of the pictorial backs were named, a number were not. Those in the latter category, while perhaps of less interest to many collectors, nevertheless are very attractive and collectible. In fact, several of the very desirable early Congress 606 decks did not have named backs although they did have named Jokers which were designed to match the card back.

The designs for this style of back were by contemporary designers and artists who were often well known to the general public, and in fact, many of the named decks were miniatures of large, well known works of art.

Undoubtedly, the best known maker of the named decks was USPC and the best known brand was Congress #606. Other makers and brands included:

- Kalamazoo Paper Box (subsequently Kalamazoo Playing Card Co.), which made Smart Set #400,

- American Bank Note, which made pictorial cards,

- Andrew Dougherty, maker of Tudor #1485, Waldorf #230 and Marguerite #130 (narrow),

- Standard Playing Card Co., maker of Society #1001, Peerless #304 and other named brands including Austen Beauties for J. I. Austen,

- Russell Playing Card Co., which made several narrow brands,

- Parker Brothers, which made both Rembrandt and Bellevue,

- American Playing Card Co., makers of Premier #550, and

- National Card Co., makers of Bijou #1.

It is surprising that NYCC, a prolific maker, did not have a really popular named brand, although they did make some pictorial backs under the Canary #911 and Triton #42 names.

The collectors of playing cards who are especially interested in collecting cards with interesting back designs typically fall into two categories – those who collect full decks of cards, including those with interesting back designs, and those who collect single card examples of various backs. Of course there are also many collectors who fall somewhere in-between, collecting both decks and singles.

The Chicago Playing Card Collectors Inc., a large club based in Chicago, has many U.S. members as well as several from other countries, including a strong group from the Australian clubs. Over a period of many years, hardworking members of this club have compiled catalogues of many different back categories, including both wide and narrow named. For this reason, no attempt has been made to further expand on this area of card collecting, or to list any of the wide variety of named cards in this Encyclopedia, as both named and unnamed decks are, of course, listed in the chapters on standard cards under their brand names. Finally, for the many collectors who are interested in these beautiful cards, we have shown examples of the brands mentioned above.

USPC ADVERTISING BOOKLET

01 – 5AM

02 – MOONFAIRY

03 – CUPIDS SECRET

04 – THE POOL

05 – BY HECK

06 – PIERRET

07 – ALPINE STREAM

08 – ROSES

09 – TROOPER

10 – AMERICAN

11 – CARNATIONS

12 – IT LISTENS GOOD

13 – TULIP

14 – GYPSY

15 – PEERLESS

16 – BOATING

17 – QUEEN

18 – UNNAMED

19 – UNNAMED

20 – THE DANCER

21 – PANSY

22 – SAPHO

23 – IRIS

24 – CORNACOPIA

25 – DEBUTANT

26 – NY GIRL

27 – DAFFODIL

28 – NANETTE

29 – DEBUTANT

30 – POINSETTIA

31 – UNNAMED

32 – MILDRED

EARLY CONGRESS AD

SINGLE ENDED CARDS

ANDREW DOUGHERTY   AD 1a

J.Y. HUMPHREYS   U29

EMPORIUM   U34

CARMICHAEL, JEWETT & WALES   U35

EARLY INDICES

SALADEE'S PATENT

NY44

TRIPLICATE

AD12

STERLING

U37

BORDER INDEX

L14

VARIATIONS ON STANDARD COURTS

PYRAMID

PY6

INTERNATIONAL

NS16

EXCELSIOR

L29a

SHAKESPEARIAN

NY54b

SPANISH-SUITED AMERICAN PLAYING CARDS

J.Y. HUMPHREYS   *U29d*

L.I. COHEN   *NY7*

COLUMBIANO #81   *SX12*

NAIPES FINOS #49   *AD25*

ILLUMINATED CARDS

L.I. COHEN *NY6*

LAWRENCE & COHEN *NY10*

ANDREW DOUGHERTY *AD3*

CARD FABRIQUE *L17a*

TRANSFORMATION CARDS

A SUPERANNUATED HORTICULTURIST.

TIFFANY HARLEQUIN

A SPRING DAY.

BACK-HAIR STUDIES.

FINNIGAN'S WAKE.

A BATCH OF BARRISTERS.

T4

HUSTLING JOE

T9

POLITICAL AND PATRIOTIC CARDS

CAFFEE

P3

PRESIDENT SUSPENDER

P4

WAR CARDS

ANHEUSER-BUSCH

W16

MLLE. FROM ARMENTIERRE

W30

INSERT CARDS

17

17a

15

112

I19

I20

12a

124

AMERICAN PINUPS

VARGAS

N35

ELVGREN

ART NOUVEAU INFLUENCES

MEDIAEVAL   *NY55*

NEW ERA   *US20a*

CONGRESS   *US6*

STAGE #65   *SE5*

ART DECO CARDS

HYCREST ROYALTY

N19

ROYAL REVELERS

GREEN SPADE

MISCELLANEOUS CATEGORIES

PHILITIS

NS25

Bid-Rite
Playing Cards

APPROACH
FORCING

SYSTEM

Bidding and Playing Values
in Contract Bridge

Copyrighted, 1932
S. R. Huntley
Chairman, Card Committee of the
Knickerbocker Whist Club

Patent No. 86134
BID-RITE PLAYING CARD CORP.
580 Fifth Avenue
NEW YORK

BID-RIGHT

BW16

NEW ERA

024

BLUE SPADE

NR6

JOKERS

SOUVENIR AND RAILWAY CARDS

S31

S9

S74

S63

S62

S1

S52

S57

SCA19

SCA28

SCA5

SCA17

SR30

SR6

SR1

SR27

AMERICAN CARD BACKS

WIDE ADVERTISING BACKS AND JOKERS

NARROW ADVERTISING BACKS AND JOKERS

MAXFIELD PARRISH

JOKERS

MISCELLANEOUS

COCA-COLA

USPC BICYCLE CARDS

STORE POSTER

TRADE CARDS

N.Y. CONSOLIDATED CARD

RUSSELL & MORGAN PTG.

VICTOR E. MAUGER

ANDREW DOUGHERTY

1893 CALENDAR PAGE

BRIDGE TALLIES

CANADIAN STANDARD PLAYING CARDS

To the best of our knowledge, nothing has been written on the subject of standard Canadian playing cards. This is not surprising as, until Gene Hochman undertook this major work, there was very little available on standard American cards, and Canada, relatively speaking, was a minor player.

We have not had the opportunity to do extensive research on Canadian playing cards, but have spent time in the public libraries in Montreal and Toronto reviewing old city directories and trade journals. From this effort and the items in our collection, we have listed in this chapter all Canadian manufacturers and brands known to us, except that categories covered in the original Encyclopedia are now in Chapter 19 (Insert cards) and Chapter 27 (Canadian Souvenir Cards).

There are a number of collectors in Canada and over the years a large number of Canadian decks have found their way south of the border and ultimately into the hands of many U. S. collectors. We therefore believe this first start at cataloguing Canadian makers will be of interest and value.

There are three distinct periods in the history of playing card use and manufacture in Canada up to the 1970s. The first of these runs up to the 1870s, and during that time it appears that all cards were imported, as there are no records easily found which indicate Canadian manufacture, despite advertisements by wholesalers and stationers for playing cards. Interestingly, one of the advertisers in 1870 was one Victor E. Mauger (refer Chapter 3) who described himself in a large ad as a "Commission Merchant" in Montreal and the depot for Goodall's Playing Cards. The second period runs from the 1870s to 1939 and the third from 1940 on.

During the last two or three decades of the 19th century several companies, located primarily in Montreal, started manufacturing playing cards. Most of these companies already were in the printing or publishing business and the addition of a playing card line seemed to make sense. At the same time, imports, primarily from Chas. Goodall in London, England, and USPC (Russell & Morgan Co.) increased dramatically. Clearly Canadians were taking up card playing in droves. In the case of some of these early manufacturers, it is difficult to know if the company shown on the Ace of Spades made the cards or whether a different printer or manufacturer made the cards for stationers and wholesalers. In any event, many of these companies did not last for too many years.

By the turn of the century, playing cards were a big business in Canada. Stationers had extensive lines of domestic and imported cards and advertised profusely. Goodall had a variety of their standard decks available, some with special backs to appeal to the local consumers, as well as many Canadian advertising and souvenir decks. The popular brands imported from USPC included Bicycle #808 and Congress #606. By this time the Canadian makers had been consolidated and only two of any consequence remained, both in Montreal.

A decision, which was to have far reaching consequence, was made by USPC in 1914 to start a Canadian operation. That year they opened a Canadian plant in Toronto and started the production of certain of their U.S. brands with a similar Ace of Spades but with a Toronto address and frequently a different Joker. In 1918 they moved to Windsor, Ontario, just south of Detroit, where they continued manufacturing (from 1933 on as International Playing Card Company Limited) until 1961. From that date until the early 1990s they processed sheets printed in Cincinnati into finished decks. At this time there are no more manufacturing facilities in Canada although a strong marketing and sales division remains.

By the 1920s USPC and Consolidated Litho of Montreal were the only makers of consequence. During the early 1920s Consolidated Litho incorporated the Canadian Playing Card Co. which took over all their brands. However CPCC struggled during the 1930s and eventually, in 1939, USPC bought out both Consolidated Litho and CPCC becoming the only maker of note through the next three decades.

Although, in many instances, examples of their cards have eluded us, we have been able to identify the following early manufacturers in Montreal:

1875   Montreal Card & Paper Co.(A. J. Auchterlonie manager), 515 LaGauchetiere St.

1880   George J. Gebhardt & Co., Craig Street

1881   Montreal Card & Paper Co. (E. Jaeger manager), 515-517 LaGauchetiere St.

1882   Burland Lithographic Company Limited, successor to G. J. Gebhardt & Co.

1883   Montreal Manufacturing Co. Limited, 509-513 LaGauchetiere St.

1884   Alain & Catelli, 1588 Notre Dame St. - factory at 40 Jacques Cartier Sq. (see factory picture)

1887   Canadian Bank Note Company - Montreal

1890   Burland Lithographic Co., 8 Latour Ave.

1890   Gebhardt Berthiame Litho & Printing Co.,
        30 St. Gabriel St.

        Montreal Playing Card Co. (Berthiame &
        Sabourin—managers), 40 Jacques Cartier
        Square

        S. Robitaille, 252 St. Paul St.

        Standard Card & Paper Co.,
        303 St. James St.

1892   Montreal Printing & Publishing Co.,
        42 Jacques Cartier Square

1892   Union Card & Paper Co., 8 Latour St.,
        successor to Burland Lithographic Co.

1897   Montreal Lithographing Co. Ltd.,
        42 Jacques Cartier Sq.

1902   Union Card & Paper Co., 252 St. Paul St.

1908   Consolidated Litho & Mfg. Co.,
        284 Parthenais St.

While the above list may be somewhat confusing, we know from the present availability of old Canadian cards that two strong playing card manufacturers emerged from this grouping – Union Card & Paper Co. and Montreal Lithographing Co. Ltd.

Union PCC appears to have been formed through a consolidation of Burland Lithographic Co., Canadian Bank Note Company (controlled by Mr. Burland) and S. Robitaille in the early 1880s. It continued manufacturing cards under the Union name until circa 1911 when it changed to Consolidated Litho & Mfg. Co. In 1923 Consolidated Litho formed a new company for the manufacture of playing cards, Canadian Playing Card Co. Limited, which continued until 1939 when it was sold to International Playing Card Company, the Canadian subsidiary of USPC.

A Union Card & Paper Co. Limited sample folder from about 1900 included the following brands: Mignonette (patience size), Imperial, Mikado, Dominion, Steamship #90, Quebec, Stadacona, Magicienne, Premier, Starlight, St. Lawrence, Good Luck, Klondyke, Sports (with sporting backs like the one pictured), Owl and Earnscliffe. Other brands made by Union included Bridge, Colonial Whist, Oak Leaf, Princess, Pyramid, Golfer #22, Royal Bridge, Snowshoes, and Whippet. Incidentally, we believe that all decks made by Union and Consolidated Litho were wide and that bridge decks were only made after the formation of the Canadian Playing Card Co.

SAMPLE CARD - SPORTS

ALAIN & CATELLI FACTORY
An interesting early photograph of a Montreal playing card factory.

CDN1

CDN1 MAPLES
Canadian Bank Note Company, Montreal, c1887. Beautiful box and back featuring maple leaves made by a company that was known for the quality of its bonds and banknotes.

CDN2

CDN2 BURLAND LITHO
Montreal, c1885. Unfortunately this deck is missing its Ace of Spades but we believe it to be made by Burland or one of its antecedents. It is of cheap construction with no indices and features interesting one way courts. The deck came with a blank and likely did not have a Joker.

CDN2a   CDN2b   CDN2c

CDN2a BURLAND LITHO
Montreal, c1887. A Steamboat quality deck with small four corner indices. It probably came with a Joker although it is missing from the one copy known (see also CDN37).

CDN2b UNION CARD & PAPER CO.
Montreal, c1890. A similar deck to the above but with Union Card replacing Burland on the Ace of Spades.

CDN2c OUR SPECIAL
Union Card & Paper Co., Montreal, c1892. Union we believe, made this deck with a generic, but otherwise identical Ace of Spades for a large stationer, Copp Clark & Co. It appears from the Joker that it was meant to compete with the Bee #92 brand made by NYCC. The name was only on the box.

CDN2d PEERLESS (not pictured)
Union Card & Paper Co., Montreal, c1892. The same deck as CDN2c but the name Peerless on the box.

CDN2e

CDN4

CDN4 ST. LAWRENCE
Union Card & Paper Co., Montreal, c1900. A great Ace that was used for many different Union decks in the early 1900s. Unfortunately we have never seen this deck with a Joker and cannot be sure it is the same as CDN3.

CDN2e UNION CARD & PAPER CO.
Montreal, c1887. This no indices deck, with the same style Ace and one way courts like CDN2, was likely made in the earliest years of Union's existence.

CDN3

CDN5

CDN3 MIKADO #125
Union Card & Paper Co., Montreal, c1895. This deck, by Union Card and Paper Co. has both a blank and the interesting Joker shown and came in a cheaply made box stamped "Mikado".

CDN5 GOLFER #22
Union Card & Paper Co., Montreal, c1908. This is probably their latest Ace of Spades and the deck has a marvelous Joker in shades of red, brown and yellow. The back is decorated with golf clubs and balls.

CDN5a

CDN5b

CDN5a GOLFER #22
Consolidated Litho, Montreal, c1911. Note that the first Consolidated Litho Ace of Spades is identical, except for the name, to CDN4. Gone, unfortunately, is the golfing Joker. The Joker commonly used with Consolidated Litho decks in the 1910s replaced it. As is common in transition from one name to another, the box says Union Card & Paper Co. despite the change to the Consolidated Litho Ace.

CDN5b GOLFER #22
Consolidated Litho, Montreal, c1917. A later version produced about 1917, with a similar Ace of Spades to CDN5a and the same Joker. All three of these decks have this identical golfing back.

Both of the Aces of Spades were used for other Consolidated Litho brands, many of them the same as the Union Card brands described previously. Newer brands included Oak Leaf and Shuffler #744 which, incidentally, described on an extra card that it was manufactured by The Copp Clark Co!

Both Union Card & Paper Co. and Consolidated Litho also made a number of interesting advertising decks. Four of these are shown. The first, c1920, is for Edison Mazda Lamps and is of interest mainly for its attractive back, the Ace of Spades and Joker being the same as CDN7. The second has the CDN5 Ace of Spades of Union Card & Paper, but in this case advertising J. B. R. & Fils (& Sons) of Montreal. What is fascinating is the back which shows two pugilists boxing in front of a very large crowd. They are clearly an American and a Frenchman – likely a representation of the Jack Dempsey/Georges Carpentier fight of July 2, 1921.

The third, dated 1926, has a variation of the CDN7 Ace of Spades with the name of a well known Montreal newspaper thereon. The back design was a copy of the front page of the newspaper for Friday, November 26, 1926 and in the bottom right corner one can see a 10 of Spades, likely part of a contest being run by the paper. The final deck is one of many produced over the years by Canada's oldest brewery, and still its largest.

AD BACKS

In 1923, Canadian Playing Card Co. Limited was formed and from that time on the playing card operations of Consolidated Litho were mainly carried on under this name, although some cards carried the Consolidated Litho name for a few years. There were only two Aces of Spades used by CPCC during its existence.

CDN6

CDN6 SPORTS JUNIOR
Union Card & Paper Co., c1900. A patience deck that came with both a hockey and lacrosse back and a hockey Joker that was used for all versions of this deck. The Ace of Spades is very similar to CDN3.

CDN6a   CDN7

CDN6a SPORTS JUNIOR
Consolidated Litho & Mfg. Co. Limited, c1915. The same deck as CDN6 but with a Consolidated Litho Ace. We have never seen this Ace of Spades on a full sized deck.

CDN7 PYRAMID
Consolidated Litho & Manfg. Co. Ltd., c1921. A newly designed and simplified Ace of Spades, in this case with the brand name thereon, was introduced about 1920. It retained the same Joker as the earlier decks. This same design was used for a large number of other brands.

CDN8

CDN8 COLONIAL BRIDGE
Canadian Playing Card Co., Montreal, c1925. The first Ace of Spades copied CDN7, substituting the new name and introducing a new Joker. This particular deck had the brand name on the Ace although in many instances the name was only shown on the box.

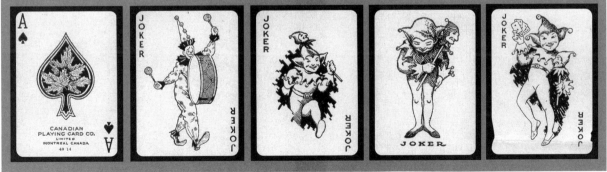

CDN9

CDN9 SPORTS BRIDGE
Canadian Playing Card Co., Montreal, c1925. The second Ace of Spades issued with the same Joker. Also pictured are four other Jokers used by CPCC. The first, a clown with drum, came in two versions, one black and white and one colored. The other three are jesters and all were used with both of the CPCC Aces.

CPCC brands, which with few exceptions were bridge size cards, included some of the older Consolidated Litho brands as well as Target, Stag, Pinochle, Sports Bridge, Devon, Tudor, Breton and Lucky. As mentioned these were often named only on the box and not on the Ace of Spades. They also manufactured a wide variety of advertising decks, sometimes with special Aces of Spades and Jokers, but often using the CDN8 and CDN9 Aces and Jokers.

From 1939 to 1952, after the acquistion by USPC, certain of the CPCC brands were continued and, although they were manufactured by IPCC in Windsor, they were shipped to customers, especially in the Province of Quebec, from Montreal.

CDN11

CDN11 STEAMBOAT #999
Toronto, 1914. This came with different backs than its American cousins but a similar Ace of Spades and Joker. A Windsor version of the same Ace of Spades is classified as CDN11a.

The United States Playing Card Company opened a plant in Toronto in 1914 for cutting and finishing operations at 249 College St., then moved to 559 College St. in 1915 as full manufacturing of decks in Canada commenced. In April 1918 they moved to Windsor, Ontario, a small city about 225 miles southwest of Toronto on the Michigan/Ontario border just south of Detroit.

A sample book from their early days in Toronto has a large number of brands for sale, but only some of them were manufactured in Canada. The brands manufactured in the United States are not listed, the following listings being for Canadian made cards only.

CDN12a

CDN12 HORNET #6
Toronto, 1914. Again a similar Ace of Spades but with a generic Joker used for many brands in Canada. Hornet also had a Windsor version listed as CDN12a. The Windsor Ace is shown.

CDN10

CDN10 INTERNATIONAL PLAYING CARD COMPANY LTD.
Windsor, c 1940. This deck was found in its original box with an IPCC seal and, based on the Ace of Spades, was clearly made shortly after the acquisition of CPCC. The Joker was the clown with drum version shown with CDN9.

CDN13

CDN13 PICKET #515
(#516x gold edges), Toronto, 1914. This deck also had the generic Canadian Joker in place of the Soldier Joker used in the United States. The Windsor version is CDN13a.

CDN14

CDN14 VOGUE #831 (#83 plain edges)
Toronto, 1914. This very rare brand in the United States was more common in Canada. It also used the generic Joker and came with at least six named backs: Opera, Chic, Tango, Bon-Bons, Auto Girl (shown here) and To The Ball Game. The Windsor version is CDN14a.

CDN16a   *CDN16b*

CDN16 TEXAN #45
Toronto, 1914. Texan was a very popular brand in Canada and continued until at least the 1980s although it had been discontinued in the U.S. around 1920. Both the Toronto and Windsor versions used the "45" Joker. The Windsor version (CDN16a) is pictured along with a later IPCC version (CDN16b)

CDN15

CDN15 BICYCLE #808
Toronto, 1914. A similar Ace of Spades and Joker to the United States version, issued with a variety of the known Bicycle backs. The Windsor version is listed as CDN15a.

CDN17

CDN17 PINOCHLE #48
Windsor, c1920. This brand was made using Bicycle #808 and National Apollo #33 backs. A later Ace from IPCC is also shown.

CDN18 LITTLE DUKE #24 *(not pictured)*
Toronto, 1914. This brand was listed in their 1914 sample book but we have never seen the Ace of Spades.

CDN15a

CDN15a BICYCLE BRIDGE #888,
Windsor, c1927. A Canadian version of the multi-color series issued in Cincinnati was manufactured in Windsor.

CDN19

CDN19 CADETS #343
Toronto, 1914. A Toronto Ace of Spades with the now familiar generic Joker.

CDN20

CDN21

CDN20 FAUNTLEROY #29
Windsor, c1932. Although the city is not shown on the Ace of Spades we believe this deck was likely made in Windsor as the box appears to be from the 1930s. It had the same generic Joker as CDN19.

CDN21 RAMBLER #22
Windsor, c1920. A National PCC brand made in Canada with USPC prominent on the Ace of Spades, with the familiar Rambler Joker.

CDN21a

CDN21a RAMBLER WHIST #22w
Windsor, c1920. This brand also came in a narrow version with a generic Ace.

CDN23

CDN23 DERBY #181
Windsor, c1930. A generic USPC brand with a different Joker.

CDN22

CDN22 HART'S SQUEEZERS #352
Windsor, c1920. The popular NYCC brand made for Canadian consumption with the Best Bower Joker. This same deck also came in a box entitled 'Great Mogul' #533, which had nothing to identify it with USPC except the tax stamp.

CDN24

CDN24 LARK #123
Windsor, c1930. A narrow generic USPC brand. Our deck is missing the Joker but, based on other decks, it was likely one of the plain generic types.

CDN24a REX #7 *(not pictured)*
Windsor, c1930. This narrow brand with a generic Ace and Joker came in a National PCC box.

CDN25

CDN25 CONGRESS #606
Windsor, c1930. While Congress #606 was a popular brand in Canada, they were not manufactured there, but imported from Cincinnati. Interestingly, many of those found in Canada came in a Congress box marked "Series B". After the wide named were phased out, the Windsor plant started the manufacture of the narrow versions. Two Jokers were designed for the Canadian market, one with a sailing ship and one with a maple leaf in place of the 'J'.

The following two brands were also listed in the 1914 sample book but we have never seen either of them, and although we can surmise that the Aces of Spades and Jokers were similar to their U. S. counterparts, we cannot be sure.

CDN26 NATIONAL APOLLO #33, *Toronto, 1914.*

CDN27 NATIONAL COLUMBIA #133, *Toronto, 1914.*

It is quite likely that there were more USPC, National and NYCC brands manufactured in Canada, but we have not come across any others, nor have we seen them advertised. Finally, we note that USPC in Canada made a huge variety of both wide and narrow advertising decks from their early days until approximately 1994.

Turning now to Montreal Lithographing Co. Ltd., we find its antecedents are more difficult to determine. It seems likely that the Montreal Card & Paper Co. and the Montreal Playing Card Co. were somehow part of the beginnings of Montreal Litho. In addition there is a strong resemblance between the one known Ace of Spades of Standard Card & Paper Co. (CDN28) and those of Montreal Litho. Further research will likely shed more light on these early companies.

CDN29

CDN29 MINIATURE
Montreal Litho, Montreal, c1897. An early deck with a generic Ace of Spades and a great Joker, positively identified by the box.

CDN30

CDN30 MONTREAL LITHOGRAPHING & PRINTING CO.
Montreal, c1900. The earliest Ace of Spades printed with the name of this company.

CDN28

CDN28 STANDARD CARD & PAPER CO.
Montreal, c1890. We have deduced that this deck was made by Standard and that the initials on the Ace of Spades stand for something like Standard Lithographing & Printing Co. Clearly the Ace of Spades is a forerunner to those of Montreal Litho which follow. The feisty Joker is unique to this deck.

CDN30a

CDN30a CANVAS-BACK
Montreal Litho, Montreal, c1905. A slightly different Ace of Spades with the "(Ltd.)" added to denote limited liability. This Joker was used at that time for all their decks. The back, showing two ducks in flight, matched the box.

These Aces of Spades are the only standard ones known for this company. Montreal Litho also made a number of advertising decks, normally using the standard Ace of Spades and Joker already shown. One in particular is listed below because of its special advertising Ace. Finally, Montreal Litho made at least one Railroad deck for the Quebec Central Railroad (Bollhagen Q2).

Other brands of Montreal Litho in this era were Cavendish, New Century, Cyclist, Boston and Extra Boston, Cardinal, Up-to-date, Steamboat, Fleur de Lis, Euchre, Frontenac and Great Mogul.

We have not seen any Montreal Litho decks manufactured subsequent to the end of World War I and can only presume that the company ceased making playing cards around that time.

CDN31

CDN31 DOMINION RUBBER SYSTEM

Montreal Litho, Montreal, c1915. This advertising deck for Dominion Rubber System has courts reminiscent of W21 and W22, the war decks also made by Montreal Litho. It has a great advertising back and ads on the face of each card. The King and Queen of each suit is a representation of King George V and Queen Mary of England and each of the four Jacks represents a different member of the armed forces. Unfortunately the Joker and box are missing from the only known deck. This deck is also listed as W22c in the war section (Chapter 20).

There are a number of other decks known, or presumed to be Canadian and they complete the listings for this section.

CDN32

CDN32 LE TRAPPEUR

Unknown maker, Montreal, c1880. These cards have a beautiful back in blue and gold and framed flesh toned courts. We have shown the colorful wrapper and Joker.

CDN33

CDN33 EMPRESS

Chas. J. Mitchell, Toronto, c1885. This deck was likely made by someone else for this publisher. It featured European style courts and heavy use of gold in the designs. The individual cards are framed on a larger card with a colorful background of flowers and birds. Tiny indices appear in the corners of the larger card.

CDN34

CDN38

CDN34 CONVERSE, COLSON & LAMB
Montreal, c1880. This deck has early two way courts and one way pips and is hand stenciled. It is possible it was made in Belgium for Canadian consumption. It has a wonderful 'beaver' back, emblematic of Canada.

CDN38 DEFIANCE #91
Copp, Clark Co. Ltd., Toronto, c1885. This deck came with two Aces, the one pictured and a generic Ace of Spades inserted loosely in the wrapper (perhaps to hide the Copp, Clark identity), and a great Joker. The quality is poor and it is not known if it was made by, or for, Copp Clark Co. Ltd.

CDN35   CDN36

CDN39

CDN35 TANDEM
Copp, Clark Co. Ltd. Toronto, c1885. There is a strong resemblance between the generic Ace of Spades on CDN38 and this Ace. We have made the presumption that this deck was made by, or for, Copp, Clark Co.

CDN36 THE PEARSALL CARD CO.
c1885. This is the only deck known by this maker. It is presumed to be Canadian although there is no conclusive evidence of this. There is no Joker or box with this deck.

CDN39 ROYAL, W, B & R
Toronto, c1900. Who is W, B & R? The letters are on both the box and the Ace of Spades and we suspect it was a Toronto stationer and that Union (see CDN5) made the deck. The deck is of superior quality and the box states "Ivory Finish". It has a unique Joker as well.

CDN37

CDN40

CDN37 BURLAND LITHO
Montreal, c1890. This deck was found very recently and could only be added to the end of this chapter. Note that it has four-way indices similar to CDN2a.

CDN40 T&B CIGARETTES
Tuckett Companies, Hamilton, Ontario, c1890. This patience size deck has no indices, a generic Ace of Spades and an advertising back and Joker. It comes in a box and is likely an advertising deck although it has some of the characteristics of an insert deck.

TRADE CARD - FRONT

MAGAZINE AD

TRADE CARD - BACK

ADVERTISING PLAYING CARDS

If every American advertising deck ever published were to be described in this Encyclopedia, the average home library would not have enough room to house the volumes. The supplements to keep pace with the annual production might even require several new volumes each year. By necessity, therefore, these decks have been listed under seven different Types, or categories, with many decks pictured as examples.

A word about value is appropriate. Advertising playing cards, like any collectible, have the values set in the marketplace. In other words, supply and demand and condition are the real determinants of value. However, two other key factors impact demand. The first is the quantity and type of advertising forming a part of the deck, together with its packaging. Aces, jokers, courts, pips and extra cards with advertising on them are desirable and the more of these special cards there are in a deck, the more desirable that deck becomes.

The second impact on value and collectibility to consider is the crossover factor. Many advertising decks have a very strong following among collectors in a particular field and this demand has a consequent effect on the value of decks. The prime example of this is Coca-Cola memorabilia where the collectors, who number in the tens of thousands, constantly search for examples of their favorite collectible, including playing cards. A Coke deck might therefore have a value of several times that of an equally nice deck, published at the same time for Stayform Corsets, a virtually unknown and somewhat less interesting collectible.

Unlike the other sections of the Encyclopedia, Gene's original plan was to have the advertising section operate simply as a guide for collectors. There are a few decks in this chapter that are featured in other chapters (e.g. advertising transformation decks) and in these cases we have given the reference number where more of the cards can be seen. We have enlarged this chapter as we feel that advertising decks have a very strong place in the American collecting field, often reflecting products and pastimes that have been long forgotten.

We have pictured examples of advertising playing cards of each of the seven Types. These were classified by Gene Hochman as Types A to G and we have changed these to Types 1 to 7. We hope this will help clarify each type of deck included and assist the reader in identifying and cataloguing advertising decks.

TYPE 1
These are listed under 'A'. This is perhaps the most desirable of the advertising categories. These decks feature many unique advertising cards, not only on the Ace of Spades and Joker but on court cards and, in some cases, on pip cards as well. Example - Murphy Varnish.

TYPE 2
'AA' listings. This category consists of decks with standard court cards, but with special messages on some or all of the pip cards as well as the Ace of Spades and Joker. Example - The Plant System.

TYPE 3

'AB' listings. For those who collect advertising cards, this category will make up the bulk of their collection. The first examples appeared in the early 1870s and a few are still being made today. A typical example will have a back design (often repeated on the box and/or wrapper) advertising the company or product and a special advertising Ace of Spades and Joker. No one will ever know how many different examples of these decks were actually produced. Example - Theo. Hamm Brewery Co.

The preceding three sections (Types 3 to 5) have been broken down into sub-categories (further into this chapter) to assist collectors in evaluating decks. As discussed earlier, the price one must pay is not only directly related to the scarcity of the particular deck, but also to the competition one finds among prospective purchasers. Products featured on playing cards such as brewery, tobacco, railroad and steamship, for example, demand premium prices. Readers will note that we have changed the sub-categories from those listed on page 26 of Volume III of the original Encyclopedia to ones we believe are more relevant. Collectors can use either group of sub-categories and can even create additional ones for their other advertising decks.

TYPE 4

These 'AC' listings have advertising on the backs but come with a standard Ace of Spades. However they must include a special advertising card, either a Joker and/or an extra card. Example – Foot, Schulze & Co..

TYPE 6

This is the most common type with advertising on the backs only. With rare exceptions they have the lowest value to the playing card collector, even though they might be very rare. For example, a small retail store might have distributed 500 plain decks with advertising on the backs and all but one of those may have been used and thrown out. The one remaining copy will still have little value to the playing card collector. A very early example of an advertising deck of this type is the steamboat deck produced by the American Playing Card Co. of Kalamazoo. It merely has an advertisement overprinted on the regular patterned back. To the ad collector it now has some added importance, but its value is likely only a little more than an early steamboat deck without the addition of advertising. Example - #1 Overprint for Household Goods; #2 Breck Shampoo.

TYPE 7 (not pictured)

An important category in the advertising field is the insert card. These are covered in detail in the Insert chapter.

TYPE 5

These decks, listed under 'ADD' have advertising on the backs and on the Ace of Spades only. It covers decks with standard jokers as well as a few early decks issued without jokers. In addition pinochle decks made for advertisers generally do not contain jokers. Example - Acorn Stoves and Ranges.

As mentioned above, Type 1 or the 'A' category, appeal most to advertising enthusiasts. In the original Encyclopedia there were 47 listings covering the period from the early 1880s to the 1970s. While there have been many decks produced in the past two decades that fit into this category, we have made only two additions for new discoveries from the 19th century. The recent productions will need to be the subject of another work.

A4

A4 CRADDOCK'S SOAP
USPC, 1895 (see SE2). A deck issued for the Eureka Soap Co. This deck was published as an advertising deck for this company with the advertising appearing on the card backs only. Each card carries a portrait photo of a stage star of the era.

A1

A1 MURPHY VARNISH
Andrew Dougherty, 1883 (see T5). This is the finest example of an advertising and transformation deck and it has a treasured place in every collection. Those who own it generally rank it among their favourites and the collectors who do not have it often place it at the top of their most wanted list. Each card is unique and designed with a Murphy Varnish product in mind.

A5

A5 CLEVELAND COMIC CAMPAIGN
A.H. Caffee, 1892 (see P3). This was one of two decks designed for the Cleveland camp in the campaign against Benjamin Harrison. All political campaign decks are considered to be advertising as that was their primary purpose. As determined subsequent to the publication of Volume III of the original Encyclopedia, there was also a deck with Harrison knocking out Cleveland (see P18).

A2

A2 KINNEY BROTHERS TOBACCO
1889 (see T8). This is the only other known full-size, advertising transformation deck. Although the pip cards are transformed and the court cards clever, only the two of diamonds makes reference to a Kinney product. Advertising also appears on the backs, the Ace of Spades, the Joker and the box. The scarcity of this deck can be attributed to the fact that it could only be obtained by saving the smaller size insert cards from the product (identical to the cards in this deck). When the complete 53 cards were saved and mailed to Kinney Brothers, the full-size deck of cards was the reward.

A6

A6 CLEVELAND CAMPAIGN
1892 (see P17). This deck was issued during the same campaign. The deck is serious and was issued to appease those party members who objected to the comic aspect of A5 above. The Ace of Spades says 'Campaign' and the court cards feature Cleveland, Mrs. Cleveland and Thurman, who was the candidate for Vice-President.

A3

A3 CROSS-CUT CIGARETTE
1882 (see SE11). This deck was issued for W. Duke Sons & Co., Durham, NC. It features advertising on the backs as well as a different message on the face of each card. In addition, each face pictures a famous stage star of that time. It is also a Triplicate deck with the card duplicated in the top left and bottom right corners.

A7

A7 PRESIDENT SUSPENDER
C.A. Edgarton Company, Shirley, MA, 1904 (see P4). This company manufactured the President brand of suspenders, a product that has all but disappeared from the market. Advertising appears on the backs, Ace of Spades, Joker, court cards and box. The double-ended court cards each feature two Presidents of the United States and their First Ladies.

A8

A8 GOLDWATER

Brown & Bigelow, 1966 (see P10). This deck was issued for the Goldwater campaign against Lyndon Johnson. The backs read AuH2O, a combination of the chemical symbols for gold (Au) and water (H2O). All the court cards, including the Queens, feature the face of Barry Goldwater. The Ace of Spades has the standard Brown & Bigelow Ace but the Jokers feature a donkey, the symbol of the opposition party, wearing a hat.

A9

A9 ANHEUSER-BUSCH SPANISH AMERICAN WAR I

Gray Lithographing Co., 1898 (see W15). Although this is a war deck, it is also a highly collectible advertising deck for the brewery collectors. The courts feature officers in the Spanish American War like Theodore Roosevelt, Admiral Dewey, etc. Advertising appears on the backs, the Ace of Spades, the Joker and the box. Every pip card has, as its background, a sepia picture of the Anheuser-Busch Brewery.

A10

A10 ANHEUSER-BUSCH SPANISH AMERICAN WAR II

USPC, 1900 (see W16). More war interest is apparent here as, in addition to the courts featuring officers, each of the fours shows a photo of a different American warship. The advertising interest is widened as, in addition to advertisements on the backs, Joker, box and background of all the spot cards (except for the fours), each of the Aces shows an ad for a different brand of beer or ale brewed by Anheuser-Busch.

A11

A11 PEP BOYS

Arrco Playing Card Co., published for the Pep Boys Auto Supply Chain. The Ace of Spades and the 'No' Joker feature the three brothers who founded the stores. The court cards are caricatures, with the Kings featuring Moe carrying an ad for Cadet Batteries. Manny is shown as the Queens that advertise Varsity Spark Plugs and Jack, appropriately, advertises Booster Motor Oil on the Jacks. The backs and box also carry advertising. An additional coupon card was added as an Extra Joker and was redeemable at any of the stores for a free 'Crying Towel' for losers. The deck was first published in 1928 as a wide deck in a slipcase and due to its success, it continued until 1940 in wide and narrow sizes as well as in pinochle and standard decks.

A12

A12 MONARCH BICYCLE

USPC, 1895. This is one of the earliest examples of an advertising deck created with special court cards. The style of these courts closely resembles the well-known Stage deck published in 1896 (see SE3). As it was issued a year earlier, it is safe to assume that the idea for this Stage deck was copied from the Monarch Bicycle deck. This is a very desirable and hard-to-find deck with special aces as well as a great Joker and extra card.

A13

A13 HAYNER DISTILLING CO.
Winter's Art Lithographing Co., 1893 (see SX11). This is an example of the conversion of a World's Fair deck to use for advertising purposes. All the cards are identical to the regular deck except for the Queen of Clubs and the Jack of Diamonds. The Jack shows a price list of the Hayner Co.

A14

A14 THE ENTERPRISE BREWING CO.
Winter's Art Lithographing Co., 1894 (see SX16). Like the Hayner deck, this one has nothing changed except two cards. In this case it is the Queen of Spades and the Jack of Diamonds which features a picture of the brewery.

A15

A15 NYCC
c1930. For want of a better name, we will call this deck the Advertising, Advertising deck. Each of the 52 cards carries the logo of a different product or the suggestion that a prospective customer might like to see their brand shown there and consequently order one of these special decks to advertise their company.

A16

A16 COCA-COLA SPOTTER
USPC Co., 1942 (see W34). The deck exists with two different backs, both of which are illustrated here. The only advertising is on the backs and the extra card (the Joker), but it is listed in this category as it was made exclusively for Coca-Cola.

A17

A17 TIME MAGAZINE
1962. A beautifully designed deck published by this weekly news magazine. Ads appear on the backs and on the box. Each pip card has a pitch for Time, many with a card game connotation. The courts and Jokers are unique and beautiful.

A18

A18 SCHERING MEDICAL
USPC Co., 1960. This deck was produced for the Schering Pharmaceutical Corp. The courts represent doctors, nurses, pharmacists and patients and each one is different. Ads appear on the backs and on the Ace of Spades. The print run was 50,000 and the decks were presented to doctors along with a dedication card.

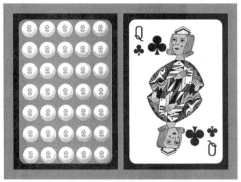

A19

A20

A19 SCHERING MEDICAL (CORICIDIN)
USPC Co., 1966. This deck, produced to promote Coricidin cold tablets, has a booklet enclosed that describes the special courts. The courts again represent doctors, nurses, pharmacists and, on the aces, patients. Unlike A18, these representations repeat on each suit.

A20 SPRINGMAID FABRICS
Arrco Playing Card Co., 1966. This deck was made for Spring Cotton Mills, a firm specializing in fabrics for household linens and ladies' undergarments. The aces, queens and two Jokers are special, being typical pinups of the times. Advertising also appears on the backs and on the box.

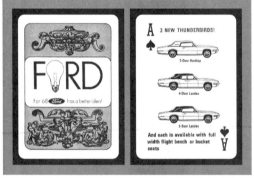

A21

A22

A21 FORD MOTOR CO.
1968. The entire 1968 Ford line is presented on the 52 cards and two Jokers. This was one of the original attempts to use a playing card deck as a type of catalogue. The backs and the box also carry ads for the company.

A22 RIDE AND WIN - FORD MOTOR CO.
1970. The backs feature the motto 'Ride and Win' plus the logos of Ford, Mercury and Lincoln. The faces have 26 sets of figures in matching pairs on the normal 52 cards, thus doubling as a child's game. The instructions are on the Jokers. All of the pictured items might be seen while touring by car.

A23

A24

A24 KENT ARMED FORCES
1970 (see W39). The backs and the box look like a pack of Kent cigarettes. No other advertising appears, but the deck is unusual and made exclusively for Kent by USPC. All cards have cartoons of U.S. Military personnel. Spades represent the Navy; Hearts - the Air Force; Clubs - the Army and Diamonds - the Marines.

A23 BEAR CARDS
Stancraft, 1968. An interesting advertising deck with cleverly designed courts made for Theodore Hamm Beer Co. Each of the courts has a cartoon of a bear dressed to participate in a different sport and features an 'H' monogram and a Hamm's Beer logo. The two baseball Jokers continue the theme.

A25

A25 STATHAM CATLOG
1954. Statham Laboratories, a Los Angeles Electronics manufacturer, issued this deck. Every card in the deck, including the two Jokers, describes a product that can be purchased from the firm.

A25

A29 JACK DANIELS
Stancraft, 1972. The cards were intended as an imitation of a 19th century deck. Advertising is on the backs, Ace of Spades and box. The cards were especially designed in a larger size (68 X 95mm) and supposedly cannot be manipulated by cardsharps. Although the imitation of an 1866 deck (the year of the company's founding) is poor, the deck is beautiful. It was not given away, but distributed for sale.

A26

A26 U.S. PLYWOOD
A modern deck which continues the latest trend of using every card, including the courts, for advertising messages. This deck, counting the three Jokers, features 55 ads.

A30

A30 BLACK VELVET
1974. Made for Heublein, Inc., Hartford, distillers of 'Black Velvet' Canadian whiskey, the deck is printed on black, similar to the Arpak deck from England. The spades and clubs are white and the hearts and diamonds, red. The only advertising appears on the Jokers, the back and the box.

A27

A27 ELLIOT ADVERTISING
c1940. This was one of the earlier advertising packs that used each card to carry a message to the public. The Joker has a picture of the Elliot Factory at Cambridge, Mass. They were well known, at the time, for a newly invented inexpensive addressing machine.

A31

A31 UNIVERSAL STARS
1941 (see SE14). Issued by Universal Studios and designed by Alexander Paal, whose initials appear on the Ace of Spades, each card has photos of prominent movie stars under contract to Universal. The Joker varies, and was used to promote various films released by Universal.

A28

A28 OFFICIAL FILMS INC.
1957 (see SE13). A special deck was produced to promote a new television series called 'Four Star Playhouse'. There is a photo of a movie star on each card with the four featured stars used on several cards.

A32

A32 LAUGH-IN
Stancraft, 1969 (see SE26). All of the court cards in this deck have the faces of cast members from this great TV comedy program. Each pip card gives several 'one liners' from one of their shows.

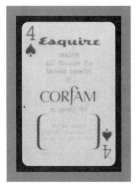

A33

A33 CORFAM
1965. This deck was issued to promote 'Corfam', a product used in the manufacture of shoes. It has specially designed courts and the pip cards and Jokers carry advertising for the company.

A36

A36 CAPTAIN CRUNCH
1976. This deck was produced for the Quaker Oats Cereal Company and created from the characters used in 'comic strip' advertisements and television cartoon commercials used to promote this new breakfast cereal. All of the cards have characters used in the advertisements.

A34

A34 REYNOLDS WRAP
1964. This deck, made for the Reynolds Aluminium Company, has a different use of the product, aluminium foil, on each card.

A37

A37 SODIUM BICARBONATE
1976. Every card in this deck for Church & Dwight Co. gives a different use for sodium bicarbonate.

A35

A35 GENERAL DYNAMICS
Brown & Bigelow, 1964. Each card has a photo or sketch that relates to the exploration of Space. The Joker pictures Superman, perhaps our first space traveler.

A38

A38 RICHARDSON BALL BEARING SKATE CO.
USPC, c1895. This rare deck is very similar to A12, the Monarch Bicycle deck. As we have never seen this deck we can only presume, from the image of the Queen of Diamonds shown, that each of the court cards promotes a ball bearing.

A39

A39 NEW AMERICAN HALF FACE EUCHRE DECK
c1880. A very unusual advertising deck from Syracuse, New York, with each card divided into two parts. One half shows advertising and the other half, the denomination. It is a euchre deck containing 32 cards and a Joker. There is only one known copy making it extremely rare.

A40

A43

A40 IBM
1969. A special deck made for employees in the shape of a computer punch card.

A43 COMPUTER AUTOMATION
c1970. The cards are the same in each suit and picture a different product on each denomination. There are a total of 16 pictured including the two jokers and an extra card.

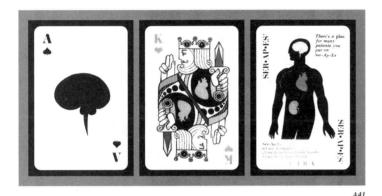

A41

A44

A41 CIBA PHARMACEUTICAL
USPC, 1969. An unusual deck made with suit signs (brains, hearts, blood vessels and stomachs) representing human parts. The aces depict the organ and the courts show where the parts are located.

A44 SCHLITZ
1970. A deck with advertising on each court for malt liquor with the kings, queens and jacks all hoisting a brew.

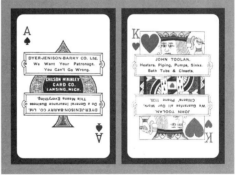

A45

A45 COMPLICATED WHIST
Kalamazoo Paper Box Co., 1905. The deck was made for Chilson McKinley Card Co., Lansing, Michigan, and each denominational card has advertising for a local company. There is a Joker as well as a 'directions' card. It could be used as a quartet game as well as for Whist and other card games. This is another very rare advertising deck.

A46 INDUSTRIAL AEROSPACE EXHIBITION (not pictured)
1964. The back of each card is different and the faces describe the backs as well as the sponsors of this exhibition held in Pinellas County, Florida.

A42

A47 GUTHMAN (not pictured)
c1970. This deck promotes the use of playing cards to advertise a company's products.

A42 PRUDENTIAL INSURANCE
1977. Each card depicts a portion of one of their advertising campaigns and the courts are stars from TV shows sponsored by Prudential.

A48

A48 FRAGRANT VANITY FAIR CIGARETTES
Victor Mauger, c1877. A very unusual advertising deck for cigarettes, especially for this early period. It was likely made by Victor Mauger (compare the Ace of Spades with Mauger's Centennial playing cards [SX1]). Each suit has a different brand of their cigarettes advertised on the cigarette smoking courts as well as on the pip cards. We do not know if this deck came with a Joker.

A49

A49 BOSTON MERCHANTS
1869. This is an advertising deck that was probably printed privately for the merchants in Boston and the surrounding towns. The cards are printed in red and black and the black and white back features a Civil War scene with horses and soldiers pulling a cannon. One of the cards shows the date 1869, making this one of the first US advertising decks known.

We now come to Type 2, covering those decks in which cards, other than the Ace of Spades, Joker and courts, are used for advertising purposes. Some use all of the aces to carry their advertising, an effective method of getting the message across, without the gigantic expense of designing new courts. Others might have advertising on all or some of the pips.

AA1 DR. RANSOM'S
c1875. This is possibly the earliest of this type of deck and the only one known without indices. Every pip card carries two remedies for a cure-all patent medicine. One card, for example, states "For Croup use both Dr. Ransom's Hive Syrup and Trask's Ointment" and "Lung and Bronchial affections cured by Trask's Ointment". Where can we find this amazing medicine today (or even the deck)?

AA1

AA2

AA2 THE GRAND MACKINAC HOTEL
c1890. This unusual deck has advertising on all of the aces, twos, fours, sixes, the Joker and the back. It was issued for the Planter's Hotel Co. of St. Louis who operated this Michigan summer resort. All of the advertising pip cards have pictures of the scenic area or hotel features. The courts are standard.

AA3

AA3 THE PLANT SYSTEM
c1890 (see example shown in the beginning of this chapter for Type 2). A similar deck, again with advertising on all of the aces, twos, fours, the Joker and the back. All of the aforementioned cards have scenes of their shipboard facilities or from their ports of call. The Ace of Spades reads "Alabama, Georgia, South Carolina, Florida, Cuba, Nova Scotia, Jamaica, and Puerto Rico." The Ace of Hearts says "Canada, Atlantic, and Plant Steamship Line." The Joker is a devil in red.

AA4

AA4 5A HORSE BLANKET
c1900. This deck, published to advertise a product that was widely used in those days, has advertising on the Ace of Spades, inside the pips on each card, on the Joker and the back.

AA5

AA5 BUCHANAN & LYLE TOBACCO
Andrew Dougherty, c1880. This deck features Triplicate type indices with advertising on the face of each card.

AA6

AA6 RCA
c1935. This deck, produced for Radio Corporation of America, has very little actual advertising. The deck is standard, except for the four aces that feature reproductions of the Sert and Brandywyn Murals from the RCA Building in N.Y. City. The Jokers describe the murals.

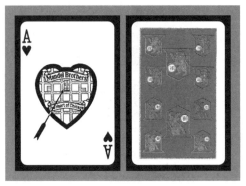

AA7

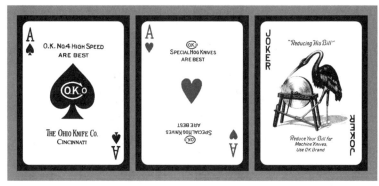

AA8

AA7 MANDEL DEPARTMENT STORE

c1910. An example of a deck with a selected card or cards carrying the message in addition to the Ace of Spades and Joker. In this case the Ace of Hearts has a map showing the exact location of this Chicago Store. The deck shown is a narrow example, exactly the same as the wide. Many companies reissued their advertising decks in narrow format to continue to interest the card player who now preferred to use the easier to handle Bridge cards.

AA9 OHIO KNIFE CO. *(not pictured)*

Brown & Bigelow, c1935. Perhaps 20 years later, the same company produced a new version of the same deck in Bridge size.

AA8 OHIO KNIFE CO.

c1915. This deck is an example of a type of deck that used one denomination card to carry different advertising messages. The Ohio Knife Co. selected the four aces.

AA10

AA10 LADISH CO.

1964. This is an example of a later deck with advertising on the four aces, showing a picture of a product produced by the company.

AA11

AA11 UNITED STATES GRAPHITE CO.

USPC, c1910. This interesting deck made for U. S. Graphite of Saginaw, Michigan, has product advertising on each of the aces and the Joker.

The following Type 3 decks feature advertising on the backs, the Ace of Spades, the Joker and often an extra card. This category is the most prolific of the early, desirable, advertising playing cards. It is impossible to picture the thousands of different decks that fall in this section, so we have pictured a few interesting examples from each sub-category. This category probably is the most sought after by collectors in other fields of collecting.

Decks listed in this AB category can be either wide or narrow. Most of the early advertising decks were wide although some narrow varieties were also issued as well in the early years of the century. By the late 1920s the wide ad decks had basically become obsolete.

No deck has been assigned a unique number. All decks have been placed in appropriate sub-categories that are grouped according to the type of product advertised. We assume that for every advertising deck, the playing card collector must compete with advertising collectors. Therefore the number of advertising collectors in each category determines scarcity, thus impacting value. We know, for example, that tobacco, liquor, brewery, automobile, and soft drink advertising bring higher prices than most of those in other fields, so we must expect to pay higher prices for cards in these sub-categories. We hope that, by using these sub-categories, it will be easier for you to group and reference your wonderful advertising cards!

AB1 TOBACCO

This is a popular and highly prized sub-category. Many have great advertising backs and aces of spades as well as jokers and extra cards. The tobacco advertising section includes cigars, cigarettes and chewing and pipe tobaccos.

AB2 LIQUOR
This sub-category includes all bottled alcoholic beverages with the exception of brewery products, the next sub-category. As is the case today, a large portion of the cost of the product in the categories of tobacco, liquor and beer was the cost of advertising.

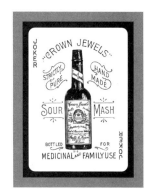

AB3 BREWERY

These decks were issued by most of the breweries in the pre-prohibition days. All brewery advertising of that period is in large demand, and although many of these decks were issued, they are difficult to find and are usually expensive.

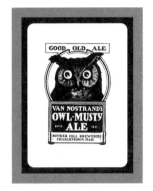

AB4 FOOD AND DRINK

Changed from Obsolete. Another well advertised category is for food and drink. Drink includes anything that is not covered in the brewery and liquor categories.

AB5 ROAD TRANSPORTATION

Changed from Automotive. This category includes means of road transportation such as cars, bicycles, buses, trucks and others. It also includes the firms that made parts and accessories for their use.

AB6 HOUSEHOLD PRODUCTS

Changed from Miscellaneous which is now AB11. There were many products advertised on playing cards to catch the eye of the busy homemaker. This section therefore includes appliances, cleaners, paints and anything else found in the home. It also includes beauty products, medications and clothing.

AB7 RAILROAD AND STEAMSHIP

This category is not to be confused with the Railroad Souvenir cards that have a different scene on the face of each card (Chapter 26). This advertising category features decks that generally have the advertising on the backs, Ace of Spades, Jokers and other extra cards.

AB8 MACHINERY AND EQUIPMENT

Changed from Specials. Another common category of advertising. It covers construction, farm, factories and other similar fields.

AB9 SPORTS

This category covers the broad range of sporting equipment, from baseball to billiards as well as decks that would promote any sporting event. An example is shown of a deck that was issued for the Roll-Crawford-Brendamour Company of Cincinnati, who were jobbers of "Athletic Equipment, Fishing Tackle, Gymnasium Supplies, etc." in 1909. The Joker honors the batting champions of 1908 in both the National and American Baseball Leagues. The league championships in baseball that year were decided on the final day of the season and one of the extra cards has the final day's box score plus the final standings of both teams (the American League on one side and the National League on the other). Another extra card has the World Series results. This deck would also be an extremely valuable item in any baseball or sports collection. We also show the W.S. Brown deck for Athletic Goods as well as two interesting sporting backs.

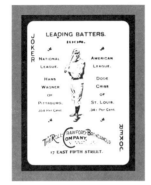

AB10 STORES
This is another large field as many of the stores carried playing cards for sale. Using them as a means for their own advertising was therefore obvious.

AB11 MISCELLANEOUS ADVERTISING

This is the largest of the AB sub-categories, including all products and services not included in AB1 to AB10. As many of the wide issues from this grouping date back to the late 1890s, you will find many decks advertising companies that are long defunct. Scarcity however, does not determine value. Presentation, oddity, originality of ideas, design and beauty will all play their part in the assessment of how desirable collectors rate a deck.

A few examples of particularly interesting narrow AB decks follow: Note their AB categories.

AB4 COCA COLA decks

Because of the large number of Coke collectors, these decks are becoming expensive and hard to find. A few of the many narrow back designs are shown along with Aces of Spades and a Joker.

AB5 S.H. HINES CO.

c1930. An advertising deck for this ambulance service. The style of the ambulance would help in dating this deck.

AB6 GENERAL ELECTRIC - EDISON

c1920. This deck has an extra card as well as the Joker that shows an early electric light bulb.

AB6 GENERAL ELECTRIC - EDISON MAZDA

1919 to 1931. Maxfield Parrish designed ten beautiful card backs for Edison Mazda. There are different Jokers and Aces in this series and each deck also comes with an extra descriptive or 'picture' card. The many Parrish collectors actively collect these decks, seeking even a single card of each back. A few of the back designs are shown in the colored section.

AB11 THE BURLESON SANITARIUM
c1920. This deck is a good example of a narrow deck that fits right into the miscellaneous AB11 section - a class in itself!

For the Type 4 to Type 6 advertising decks, we will continue to follow the numbers used in the AB category. The numbers, which follow the letters, will always represent the same sub-category, e.g. AB1, AC1 and ADD1, will always be tobacco advertising, the number '1' being allocated to the tobacco decks. AB2, AC2 and ADD2 will always be liquor advertising, AB3, AC3 and AD3 will always be breweries, and so on.

Type 4, the AC category, represents those advertising decks in which there are no cards changed. The Ace of Spades and all courts and pip cards are standard. The advertising appears only on the backs, on the special Joker and/or perhaps on an extra card or two. There is usually a special box. The following are examples of this type.

AC2 HUNTER BALTIMORE RYE
Samuel Hart, c1905. A standard NY50 Ace of Spades comes in this deck that includes two extra cards for Wm. Lanahan & Son, Baltimore.

AC6 COCOA NAPTHA SOAP
Andrew Dougherty, c1900. This deck, with the colorful Joker, clearly comes under the category for Household Products.

AC 11 DR. A.C. DANIELS - VETERINARY MEDICINE
The Kalamazoo Playing Card Co., c1900. This deck has a standard Kalamazoo Ace of Spades, a special Joker, advertising backs and an interesting box.

AC11 THE BOSTON HERALD
Andrew Dougherty, c1895. (Note the change from AC6, now Household Products, to the Miscellaneous AC11 category). This deck has a standard Ace but comes with a special early Joker. It also comes in a version with a special Ace of Spades.

Type 5 decks have advertising on the backs and Aces of Spades only. They are usually very early decks or from the 1920s on. 'AD' was originally used in Volume III but changed to 'ADD' so as not to repeat the use of 'AD' for Andrew Dougherty in Volume IV.

ADD6 FRIGIDAIRE

c1928. This deck has an advertising Ace of Spades and a very colorful back. Note the standard Joker. It obviously belongs in Household Products.

Type 6 decks are the most common of all of the advertising cards with advertising on the backs only. Decks have been made for hundreds of thousands of companies and organizations, particularly in the years after WWII. Most of these decks have little value to playing card collectors. Examples are pictured under Type 6 earlier in the chapter.

Finally, Gene Hochman had two other small categories for advertising decks. These were the 'AE' category for giant size decks (4" x 6") and 'AF' for decks which came in unique advertising packaging. We have not pictured these, but brief descriptions follow.

In the 'AE' category an unusual oversized deck was issued in 1972 for the University of California in Riverside. On all of the backs, plus some of the faces, there was information about the school as well as their courses.

Another deck of this type was made for the Chamberlin Metal Weather Strip Co., of Detroit, Michigan in 1927. All of the faces are standard with only the Joker giving an advertising message. However, each of the 53 backs illustrates different buildings that used their weather stripping and also shows the location, the date completed, the architect and the contractor.

His other category was 'AF' where the advertising deck or decks come in packaging that was especially designed for the product. Two examples include a special leather case for Michelin Tires with their logo printed in gold on the front and a marvelous Bakelite box for Gamewell Fire and Police Alarms. The box is shaped and designed like a fire alarm box and contains two decks of their advertising cards.

TRANSFORMATION CARDS

A new collector will usually start by trying to accumulate as many items as possible in a particular field of interest. Sometimes this is not wise, as there can be two obstacles to collecting this way, especially when the field is as vast as that of playing cards. It is often very difficult to find scarce items and it can also be a problem to come up with the funds needed to buy a special object when it is found.

Specializing often results when the collector has finally faced the second obstacle. In order to build a fine collection it becomes necessary to resist some items in order to have the available capital to purchase those more desirable ones.

Gene Hochman had always maintained that transformation decks were his favorite, although he also was fascinated by tobacco insert playing cards, and so when he found one he would often have to sell some other treasure in order to own that special deck. When it happened the first time little did he realize that he was on his way to specializing.

What is a transformation card? It is a card where the pips (suit symbols) are used as part of a picture or design on the face of that card, often comical. To put it another way, the pip cards have been 'transformed' by the addition of the clever drawings incorporating the pips. Transformation decks possibly have the greatest following in the world of playing card collecting. Many collectors put them at the top of their list and with good reason, for they are wonderful, whimsical, and probably the most fascinating of all the many varieties of playing cards.

Let us turn back the clock to the early 19th century. Many cards of this period remain as examples of hand-drawn transformations, often beautifully executed in watercolors or with pen and ink. Playing cards during that period had no corner indices, so an Ace of Diamonds, a deuce of Hearts, or any pip card might suggest an idea to wile away an hour of doodling. Thus the name, the transforming of a simple card into an amusing work of art.

Although earlier examples of printed individual transformation cards, as well as printed sheets, are known, probably the first completed pack is attributed to J.G. Cotta, Tubingen, Germany. The Cotta deck was known as Die Spielkarten Almanach as it was sold and packaged with a descriptive booklet and a calendar. The first was printed in 1804 for distribution on New Year's Day, 1805. This deck was in honor of Joan of Arc. Five more beautiful artistic decks were issued by Cotta, the last one being published in 1811.

These whimsical decks soon caught on and other transformation packs were issued. Throughout the world, until recent years, only about 70 different transformation decks appear to have been issued. Many were similar to each other, with slight variances when produced by an another manufacturer in a different country. It was not until 1833 that the first American transformation deck appeared.

In the original Encyclopedia Gene Hochman provided pictures of most of the individual transformation cards. For space reasons, and because a book on the subject (*Transformation Playing Cards* published in 1987 by U.S. Games Systems, Inc.) shows pictures of each card, we only show examples of each deck in this chapter.

T1

T1 BARTLETT TRANSFORMATION
Charles Bartlett, NY, 1833. Although this was the first American transformation deck, it was not original. It was nearly identical to three decks produced in various parts of Europe about 15 years earlier. The deck was known in each case as Beatrice or The Fracas, a Viennese story that was very popular at the time. This was a misnomer as only four cards of the Spade suit (3, 7, 8 and 10) were actually scenes from this story. The court cards were in classical costume.

The American deck was almost an exact copy of the Ackermann version. It is easy to distinguish it from the others as all of cards are gaudy and crudely colored. Purples, bright greens, reds, etc. give the deck a startling effect and detract from the beauty.

The original was believed to be by Rudolph Ackermann and appeared four at a time, in the Repository of Arts, an English magazine, during 1818 and 1819. The court cards had backgrounds and were delicately colored, giving the deck a fine artistic look.

The fact that Beatrice was a Viennese story leads some to believe that the H.F. Muller deck made in Vienna (1818 or 1819) was the original. It seems difficult to think that the highly regarded Repository of Arts would publish a copy without crediting the originator. This deck has no background on the court cards which were also beautifully colored.

The third issue was made by Gide Fils in Paris c1820. It is a beautiful deck with the shading and engraving somewhat improved. Rapid identification can be made, as the courts, although without backgrounds (like Muller), are standing on platforms.

T2

T2 SAMUEL HART TRANSFORMATION

Samuel Hart & Co., NY and Philadelphia, c1860. This deck is copied from one published by Braun and Schneider, Munich, Germany ten years earlier. However it is not colored and has the Samuel Hart name on the Ace of Spades. The pip cards are in brown and the court cards read King, Queen or Jack.

T2a

T2a SAMUEL HART TRANSFORMATION

Samuel Hart & Co., NY and Philadelphia, c1860. There is a variation of the deck where the Hart name is removed from the Ace and Jack of Spades but left (mistakenly?) on the Queen of Clubs (see close up). In addition the Ace of Hearts was changed to be more acceptable!

T3

T3 ECLIPSE COMIC PLAYING CARDS

F.H. Lowerre, NY, 1876. This deck was the first original transformation deck to be published in the United States. The Ace of Spades is striking and bold with the name of the deck printed on it. The comical court cards are framed in gold and the crowns of the Kings and Queens are also gold. Gold accents appear throughout the deck making it striking as well as unusual. Despite its beauty, it has, in the view of many, some faults, being not as artistic as many of its European predecessors. The themes are not as clever and there are many suit signs that are unused in the overall design. However, it is a rare deck and a gem in any collection. This is the first transformation deck to be issued with a Joker.

T4

T4 TIFFANY HARLEQUIN PLAYING CARDS

Tiffany & Company, NY, 1879. This delightful deck, designed by C.E. Carryl is, without question, the most skillful and artistic of the American transformation decks. The pip cards have clever witticisms and the court cards are modified in a humorous manner. Tiffany thought so highly of these courts, that they had them reprinted in a deck that was published in 1974.

T5

T5 MURPHY VARNISH TRANSFORMATION

A. Dougherty, NY, 1883. This deck is a transformation deck and an advertising deck that was made for the Murphy Varnish Company. Playing card collectors covet this deck as the transformed pips contain comical slogans referring to the company and its products and the attractive brightly colored courts also advertise Murphy Varnish.

T6

T6 HARLEQUIN INSERT PLAYING CARDS

Kinney Tobacco Co., New York, Richmond, Baltimore and Danville, 1888 (refer also I8). This is a tobacco insert deck with each miniature card packed individually in a package of Sweet Caporal cigarettes. Complete decks are rare as Sweet Caporal smokers had to collect the cards one at a time in order to amass a complete deck. The cards were an exact copy of the Tiffany Harlequin Deck, except that many of the designs were reversed. Also the Ace of Spades was changed to advertise Kinney. The backs of the cards listed the 52 Harlequin cards so that smokers were aware that a full pack could be accumulated.

T9

T7

T7 HARLEQUIN INSERT PLAYING CARDS

Series 2, Kinney Bros., 1889 (refer also I9). The original insert card campaign was such a success that Kinney followed up with a completely new transformation deck. Any smoker who was able to collect all 53 insert cards could send away for the full size transformation deck (see T8) as a premium. These two decks had identical faces, their only difference was size. A complete listing of the available cards was on the back of each insert card and the deck included a Joker.

T9 HUSTLING JOE I

USPC, 1895. This was the first entry into the transformation field by USPC. However, this deck is not really a transformation deck in the true sense of the word. However it is clever, tells a story and vaguely transforms some of the pip cards. The Ace of Spades features Hustling Joe dressed in red and the Joker is a rather devilish fellow in orange.

T10

T8

T8 KINNEY TOBACCO TRANSFORMATION

Kinney Bros., 1889. This full size Kinney deck was new and cleverly transformed. Every pip on every card was utilized in the designs, including those of the court cards. This is the first American transformation deck to include the courts in transforming the pip designs.

T10 HUSTLING JOE II

USPC, 1895. This reissue was still dated 1895. The reason for the reissue was the discovery of a mistake that went unnoticed until the deck was being used. As the court cards had colored backgrounds and each suit was a different color, and all of the spot cards had white edges, a sharp player might be able to note the honor cards when they were being dealt. This made the deck suspect when playing card games. The reissued pack had courts with the different colored backgrounds surrounded by an irregular white border making them identical to the pip cards when viewed from the edges. On this deck, Hustling Joe wears yellow on the Ace of Spades.

T11 VANITY FAIR
USPC, 1895. America's largest playing card manufacturer finally made a true transformation deck. In this edition all of the spot cards are cleverly used in imaginative designs. The court cards, while not transformed, are comical, quite similar in feeling to the Tiffany deck.

T12 YE WITCHES' FORTUNE-TELLING CARDS
USPC, 1896. As this deck is not really a transformation deck it is completely described in the fortune-telling category (FT9).

In the mid-nineteenth century, a completely new type of transformation card was published. The individual pips, instead of being incorporated into a design, were used as the background for faces or figures to be drawn within each pip. The first of these was designed by A. Crowquill, and published by Reynolds & Sons, London, c1850. Shortly after, in 1863, a subsequent deck was issued by C.B. Reynolds of Liverpool to honor the wedding of the Prince of Wales (later Edward VII). The Crowquill cards were printed in black and white. To distinguish suit colors, the Spades and Clubs were printed on a blue background while the Hearts and Diamonds were done on buff. The C.B. Reynolds (no connection to Reynolds & Sons) deck was printed in the standard red or black.

T13 FUNNY SPOT
Continental PC Co., NY, 1905. The only American issue of this type was done after the turn of the century. The pips had humorous faces and, in addition, every card contained a motto or humorous saying. Again, these cards are not true transformations, but as this clever deck merits recognition this is certainly the right category to put it in.

T14 SUTHERLAND-BROWN
Laura Sutherland, 1977. While not a true transformation, this deck is included here as the pips were used as the base for the drawings.

In recent years a number of transformation, and near transformation, decks have been designed and published. Some of these have been issued in the United States but have not been included here due to their relatively recent publication.

INSERT CARDS

Starting about 1870, a completely new idea for advertising originated in the United States. This happened to coincide with the addition of color on the printed advertising card. Aiming at the collecting urge in the average person, series of different trading cards were inserted in products that were marketed in small packages. The subjects that were used on these cards varied. They included baseball players, flags, racehorses, statesmen, Indians, soldiers and birds, to name just a few. The variety of advertised products varied almost as much as the cards, and included gum, candy, cereals and, most notably, tobacco. The first playing card inserts were used about 1885. What better way could there be to persuade people to buy a product than the need to collect and fill in a certain number of cards to complete a series? Therefore playing cards, which numbered 52, became a natural for inserts.

The insert card trend rapidly caught on around the world. Playing card inserts were published everywhere: England, Germany, France, Holland, Venezuela, Peru, India, Africa, Cuba, Chile, Canada, China and many other countries as well. Especially noteworthy were several beautiful editions inserted into chocolate boxes sold in Spain.

Gene Hochman was fascinated by tobacco insert playing cards and found it fun to complete series one by one just as done originally. One often finds groups of these cards that do not belong together or are incomplete. So often, many years before, some smoker casually collected them just as trading sports cards, etc. are collected today. Many collectors have lists of cards they are missing in the hope that one day they will accumulate a whole deck of insert cards.

There were a number of different editions issued at the turn of the century by the American Tobacco Co. They were not distributed with any product sold in the United States. They were issued first by ATC, then by the British-American Tobacco Co. (BAT) in Great Britain, with Ogden's Ltd., Cameo, Cross Cut, Vanity Fair, Old Judge, Cycle, and Pinhead brands.

In this chapter we have pictured the back of each listing and a few examples of the faces. Space constraints have not allowed us to picture each card as was done in the original Encyclopedia for many decks.

11

I1 LORILLARD'S 5 CENTS ANTE CHEWING TOBACCO
P. Lorillard Co., NJ, c1885. These first US playing card inserts were small, 3 1/2 x 1 3/4 inches. The cards were issued in an edition of 52 cards and featured beautiful girls of the day dressed in theatrical costumes. They were printed by Donaldson Brothers, Lithographers, NY.

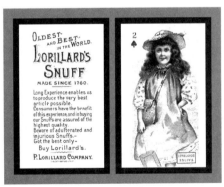

12

I2 LORILLARD'S SNUFF
c1885. This rarer edition has the same pictures as I1 but the backs advertise Snuff and the logo on the faces says 'Lorillard's Snuff' in place of '5 cents Ante'. The cards are the same size. Both series had 52 cards and no Joker.

12a

I2a LORILLARD'S SNUFF
c1885. This scarce 'transition' version has the I2 back overprinted with the '5 cents chewing tobacco' information from I1.

13

I3 HARD A PORT BLACK BACKS
H.A. Thomas & Wylie, Lithographers, NY, c1886. This is one of three issues using identical pictures and printed by the same company. The backs are a very dark green that gives the appearance of being black. They are 2 3/8 x 3 7/8 which is larger than most other insert cards. The phrase "Hard a Port Cut Plug" appears on the face of each card. I3 and I4 were sometimes issued without a Joker while I5 was issued as 52 plus Joker.

14

I4 HARD A PORT BLACK BACKS II
c1886. The identical pack as I3 with the identical backs. The only difference in the cards is that the brand name is not shown on the faces of the cards.

15

I5 TRUMPS LONG CUT BLACK BACKS
c1886. The date of the three packs in this series is approximate as it is not known whether they were printed simultaneous or at different times. The pictures on this edition are identical to I3 and I4, but the faces show the 'Trumps Long Cut' logo. This deck had 52 cards plus a Joker.

16

I6 KIDS PLUG CUT
Gravely & Miller Tobacco Co., Danville, VA. c1887. This unusual edition was one of several put out by this company. They used the same face designs, only changing the backs and face logos. The series differs from most of the others as the designs are repeated in each suit. There are a total of 14 poses, one each for the Ace to the King and one for the Joker.

17

I7 SNIPE PLUG CUT
Gravely & Miller, c1887. Another issue which was identical to 'Kids' but with the backs and face logos changed. Besides being able to collect an entire pack by using these brands, they were also offered as a premium for 50 tin tags that came with any of the company's brands. In spite of this, these insert cards are very scarce and complete sets are a rarity.

I7a RED BOOT LONG CUT
Gravely & Miller, c1887. This is yet another identical, very rare, issue with only the references changed.

I7b DEEP RUN HUNT CLUB WHISKEY
Gravely & Miller, c1887. This is the first and only known insert deck (again very rare) of this period that is not a tobacco deck. It is still another one of these identical insert cards except for the references to the product.

I8 HARLEQUIN INSERT PLAYING CARDS
Kinney Tobacco, 1888. The insert version of T6 described in the transformation chapter.

I9 HARLEQUIN INSERT PLAYING CARDS
Series II, Kinney Tobacco, 1888. The insert version of T7 described in the transformation chapter.

I10 KINNEY TRANSPARENT CARDS
c1890. An interesting deck of transparent playing cards which when held to the light reveal a picture and a 'fortune'. The first issue comprised 52 cards without a Joker. The backs read "Transparent Cards" at the top and the cards are 1 1/2 x 2 3/4 inches.

I11 KINNEY TRANSPARENT CARDS
c1892. The backs were slightly changed and a Joker was added to make a 53 card deck. This was so stated on the backs.

I11a KINNEY TRANSPARENT CARDS *(not pictured)*
c1892. A second type of back (not pictured) was issued with this deck on which the lead line was "Read your fortune".

I12 TRUMPS LONG CUT BROWN BACKS
c1890. One of three editions that were made for the tobacco company which produced this brand as well as the Hard A Port brand. All three editions were designed to promote both of these brands by using the various backs and logos. The same face designs were used for each of the three different backs. The brown back featured a man in the center circle holding a hand of cards.

I13

I13 HARD A PORT BLUE BACKS
c1890. The same face designs appear with a 'Hard A Port Cut Plug' logo. The interior of the circle has a seaman at the wheel and the circle is 1 1/4 inches.

I14 HARD A PORT BLUE BACKS *(not pictured)*
This edition is identical to I13 except that there is no brand name appearing on the face of the cards. This seems to be the rarest of the three varieties.

I15

I15 MOORE & CALVI
Lindner, Eddy & Clauss, Lith., NY, c1890. This variety of tobacco insert card is the most common of all the types. There were no less than six editions, all printed by the same lithographers and they all advertised Hard A Port and/or Trumps Long Cut. Moore and Calvi, the original producers of these brands issued the first. The logos on the faces read "Hard A Port".

I15a

I15a MOORE & CALVI
Lindner, Eddy & Clauss, Lith., NY, c1890. This deck was also issued with the name "Cullingworth" overprinted in red on the backs.

I15b

I15b MOORE & CALVI
Lindner, Eddy & Clauss, Lith., NY, c1890. Another version has the MacLin-Zimmer overprint on the backs.

I16

I16 MACLIN-ZIMMER
c1890. The identical cards, with the same faces and logos as I15, except that the backs now carry the names MacLin and Zimmer, the successors to Moore & Calvi. Many collectors have complete decks with slightly different backs because there were so many different variations.

I17

I17 MACLIN-ZIMMER-MCGILL
c1890. Shortly afterwards the above company added a partner and used the same cards with another back. All of the backs mentioned so far feature the name 'Wake Up Cut Plug' but this brand name was never written on the faces.

118

I18 MACLIN-ZIMMER-MCGILL II

c1890. The identical back to I17 but the faces read "Trumps Long Cut" in addition to the 'Hard A Port' logo. This variety is the rarest of these six issues.

119

I19 HARD A PORT BLUE BACKS

c1890. The same overall back design, with a seaman at the wheel and the 'Hard a Port' brand replacing the brands and company names used in the above four types. This is the most common of all of these issues. It is believed that these were used as a premium given for tin tobacco tags as well as being used as insert cards. This type comes in its own case also with the Hard a Port name. The circle is larger than the similar one used for I13 and I14, being 1 5/8 inches.

120

I20 TRUMPS LONG CUT BLUE BACKS

c1890. This is the last of the six variations. These 'Trumps Long Cut' cards are just slightly wider than the preceding five. They measure 2 3/8 x 3 7/8 inches instead of 2 1/4 x 3 7/8 inches. Note that the spot cards in all six varieties have pips in the designs of their costumes.

121

I21 W. DUKE, SONS & CO. SERIES I

c1888. These insert cards were produced to advertise 'Turkish Cross Cut' cigarettes. The backs have an overall pattern and the faces are standard with no indices. Only the Ace of Spades and the Joker mention the product. They measure 1 1/2 x 2 3/4 inches.

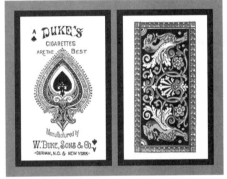

122

I22 W. DUKE, SONS & CO. SERIES II

c1890. This slightly later issue was made to promote Duke's cigarettes. The backs and faces were very similar to I21 but indices were added. The Joker for each issue was the same and can only be verified by examining the nearly indistinguishable variations in the dots on the backs.

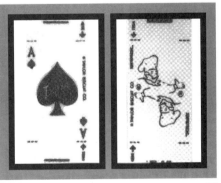

123

I23 TAYLOR BISCUIT CARDS

1959. This unusual series is only 40 years old. The cards measure 1 3/4 x 2 1/2 inches and were packaged with the biscuits.

I27 ADAMS GUM CO. CARDS
Two strips of cards (15mm x 20mm), ten cards in total, were packed in each package of Adam's Pepsin or Tutti Frutti gum. One strip had blue backs and the other red backs. When connected, the blue backs read: "Here are two complete hands. Which wins is still the problem. Break them apart and play them with your friend. Clubs are trumps and so is Adams Pepsin Tutti Frutti". The red backs read "All Bicyclists chew Adams Pepsin Tutti Frutti. Keeps mouth and throat moist the teeth clean and digestion good". The five cards in each hand were Euchre hands.

127

I24 WM. S. KIMBALL & CO.
Printed by Julius Bien Lithographers, Rochester, NY, c1895. Kimball's were the manufacturers of Old Gold Cigarettes and Vanity Fair Tobacco. These cards were used for both American and foreign distribution. The American Tobacco Company used the same series of pictures in Great Britain. The cards were also of the smaller variety.

CANADIAN INSERT CARDS

Research into the types of playing card inserts issued in Canada has revealed the following: 1) A great many from W. C. MacDonald Co. of Toronto, makers of British Consol Cigarettes, Skyways Pipe Tobacco, Brier Tobacco, Zig Zag Cigarette paper, etc., 2) at least 2 sets by D. Ritchie & Co. of Montreal, c1895, 3) at least 2 sets by Imperial Cigarette Co., C. P. Mitchell, agent, 4) a set by Imperial Tobacco Co. of Newfoundland, 1930, and 5) a set by Quaker Candy Co. We have not been able to obtain pictures of the Imperial Cigarette or Imperial Tobacco cards.

W. C. MacDonald Co. issued playing card inserts in their products. They were redeemable for one full size deck of playing cards if a full deck of inserts (52 cards and the Joker) was collected. The promotion was started in 1925 and was so successful it was continued until 1948. The cards had an expiration date and those who attempted to collect a set for each date would require 90 decks to complete the series. The company continued to change the design of the cards (there were 40 designs in all) in addition to changing the dates to help the checkers who were examining the redeemed sets.

The 60x41mm cards issued between 1926 and 1947 by MacDonald's come in 29 basic designs incorporated into 84 separate issues with numerous variations of print, color, format and paper quality. The first MacDonald cards have one design that is an illustration of the factory and is inscribed "The Home of MacDonald's Tobacco and Cigarettes".

I25 WOOL CANDY
c1935. Technically, this deck should not be listed here as the cards were sold in candy shops in strips of 13 (a complete suit) for 1 cent. They bear the name of the company on the back of each card, along with the identity of the card.

The next four printings, issued in 1927, used as many as eight designs. All issues after the first five were printed with a serial number inscribed near the playing card value, and, except for the first 11 serial numbered printings, are numbered 1 to 53. The Joker is #53 and in most of the later printings has a "Score Memo" on the back. This Joker depicts an early Canadian hockey player.

I26 PREMIUM SLIPS
Copyright 1931, Pat. Pending. Long, narrow white paper slips 5/8 x 2 3/4 inches. The black suits were printed in navy blue and the red suits in red (both sides). The backs listed the premiums that could be had for collecting. For a complete deck (52+J), a Bicycle; for 4 Aces, a Target Rifle; for 4 Kings, a man's wristwatch, etc. Gene Hochman stated that he had 46 of the 53 cards and knew three other collectors with many cards, yet not one premium could be claimed through combining the collections. One can speculate as to how many of his missing cards were actually issued.

126

Seemingly, the scarcest design is titled "National Hockey League Games 1934-35" which lists various game dates and scores. The most popular and plentiful cards are the Aeroplanes and Warships issued during WW II. With these tobacco cards it is possible to collect a 52 card deck with the same ship or plane back or a number of different backs with the same expiry date or any combination of date, ship or aeroplane.

ICA1

ICA1
The first series was released in 1925. They cards are white with at least six different advertisements on the backs. The expiration dates are March 31, 1926 and March 31, 1927.

ICA2

ICA2
This series, with the year overprinted in large numbers, started with the expiration date March, 1927. A new face design was then introduced which included the Brier name. This series had expiration dates running from March 31, 1927 until March 31, 1931. All expiration dates except for September 30, 1927 and December 31, 1927 were printed in black with a red overprint. Those two dates were printed in red with a black overprint. Most of the backs had the same advertising as ICA1. One new back offered additional premiums for collecting more than one set. Some of the remaining ICAl cards were also overprinted in this manner.

ICA3

ICA3
Large overprinted letters replaced the overprinted numbers from expiration date May 1932 until September 1936. The faces of the cards were the same except for the addition of a number added in the lower left hand corner. The cards were numbered from 1 to 53.

ICA4 (not pictured)
A special 'Blue Chip' series had an August 1932 expiration date. It was printed on a blue card offering three decks instead of the usual one. No overprinted letter was used.

ICA5

ICA5
In 1937 the cards were changed to yellow. The backs of the first series featured a motto of the clan McDonald and there was a new design on the faces. The expiration dates were March, June, and September 1937.

ICA5a

ICA5a
In December 1937, and continuing until the expiration date of September 1939, the copy was changed back again to include 'British Consols' and 'Brier' at the top. The card was yellow and the backs featured a girl in a McDonald Tartan kilt. This back was introduced in the ICA3 series.

ICA6

ICA6
The expiration date of December 1939 saw the cards once again changed to white. The backs of cards had a banner reading 'Daily Mail' with a line sketch of an airplane. From XD Mar. 1940 until Mar. 1943, the sketch was replaced with a photo of one of more than 80 different planes (four different series in total). All cards had the 'Daily Mail' banner plus the name of the plane. There were many overlapping expiry dates with ICA6a, ICA6b, and ICA7.

ICA6a *(not pictured)*
The same 'Daily Mail' backs had blue faces during XD Mar. 1941.

ICA9

ICA6b

ICA6b
During 1941 and 1942, some backs had a special offer to those civilians who wished to send cigarettes to Canadian Service men overseas. These were issued concurrently with all of these variations.

ICA9 QUAKER CANDY
Toronto, c1945. Considering the rather recent estimated dating of the card, it is surprising that it is the only one we have seen.

ICA7

ICA7
From XD 1943, 'Skyways Pipe Tobacco' replaced 'Daily Mail' as an advertisement which accompanied a new series of airplanes.

ICA10

ICA10 D. RITCHIE & CO.
Montreal, c1895. The only early Canadian insert card producer made two different series of playing card related inserts.

ICA8

ICA8
From XD 1942, and until XD June 1948, the last in this long running series, war ships were shown on the backs with the name of the ship printed thereon. There were four series with warships and they ran, in some cases, concurrently with the others. No cards of any type were honored after June of 1948.

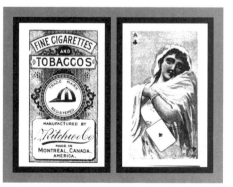

ICA10a

ICA10a D. RITCHIE & CO.
Montreal, c1895. The second version has a different back as well as faces. The Ritchie cards are very rare.

WAR CARDS

War has been a subject of fascination for generations and many interesting early decks portray battles and wars. Others were issued to honor national heroes. And, of course, the standard patterned kings and jacks, that originated hundreds of years ago, still carry weapons in our modern decks.

Research has shown that card playing was, and perhaps still is, the number one recreation of members of the Armed Forces. What better way is there for troops to occupy their minds between battles? On any ship or submarine a card game would be continually in session for the off duty members of the crew.

The early war cards are rare and valuable. They are ardently sought by military collectors, history buffs, museums and nostalgia fans as well as by playing card collectors. As always, the crossover interest makes the search for these highly desirable cards difficult and it is a thrill when we are able to add a new deck to our collection. Decks with face interest are the most eagerly sought after but this chapter will include those with war backs and other related features. Narrow as well as wide examples will be included and the chapter tries to cover all decks issued to World War II.

W1

W1 DECATUR CARDS
Jazaniah Ford, Milton, Mass, 1815. This early war deck was manufactured by one of the earliest documented American cardmakers. It features specially designed court cards which commemorate the Tripoli War. The Ace of Spades features the USS United States commanded by Stephen Decatur whom the cards honored. As some of the courts are in Eastern dress it is probable they refer to his Algerian exploits.

W2a SEMINOLE WARS *(not pictured)*
c1840. This deck was reproduced and issued by Abbot & Ely, successors to Caleb Bartlett around 1835 to 1840.

W2b SEMINOLE WARS *(not pictured)*
Abbot & Ely, c1970. A reproduction of the Abbot & Ely version was made for Old Sturbridge Village in Massachusetts in the 1970s (see NR3).

W2

W3

W2 SEMINOLE WARS
J.Y. Humphreys, Philadelphia, 1819. A beautiful hand-colored deck with stenciled pips. This was the first American no-revoke deck. The suits are in four colors: blue (spades), red (hearts), yellow (diamonds) and green (clubs). The Jacks feature Indian Chiefs and the Kings represent Washington, Jefferson, General Andrew Jackson (the Commander of the War) and John Quincy Adams (the Secretary of State who ordered the invasion of Florida).

W3 LAFAYETTE CARDS
Jazaniah Ford, Milton, Mass, 1824. This deck was issued to honor the Marquis de Lafayette on his return to the United States. The same court cards featured on the Decatur cards are repeated although new plates were obviously used as the subjects are in finer detail. The Ace of Spades has a beautiful hand colored portrait of Lafayette, who was the French hero of the American Revolution.

W4

W4 MEXICAN WAR CARDS

c1849. Most collectors agree that this deck is German although the titles are in English and there is a possibility that it may be American. It is a beautiful, hand-colored deck with each Ace depicting two battles. The Kings represent American Generals; Spades - Gen. Wm. J. Worth, Hearts - Major Gen. Winfield Scott, Diamonds - Major Gen. Zachary Taylor and Clubs - Gen. D. Twiggs.

W4a MEXICAN WAR CARDS II *(not pictured)*

Belgium, c1870. A second edition of these cards, probably made by Brepols or Van Genechten. The battle scenes on the Aces are identical but it has standard German courts.

W5

W5 UNION PLAYING CARDS

American Card Co., New York, 1862. This deck is the most common as well as the most written about Civil War deck. The box reads: "The American Card Co., confident that the introduction of National emblems in the place of foreign, in Playing Cards, will be hailed with delight by the American People, takes pleasure in presenting the Union Playing Cards in the fullest confidence that the time is not far distant when they will be the leading card in the American market. The Union Cards are calculated to play all of the games for which old style Playing Cards are used.

The suits are Eagles, Shields, Stars and Flags. Goddesses of Liberty replace the Queens, Colonels for Kings, and Majors for Jacks. In playing with these cards they are to be called by the names the emblems represent, and as the emblems are as familiar as household words, everywhere among the American people, they can be used as readily on the first occasion as cards bearing foreign emblems."

W6

W6 UNION PLAYING CARDS II

American Card Co., New York, 1863. This is a slightly later, much scarcer, edition of W5. The court cards have no background.

W6a

W6a UNION PLAYING CARDS II

American Card Co., New York, 1863. A variation of W6 with Canteen Girls replacing the Goddesses. Each holds a wooden cask and a partially empty glass of wine!

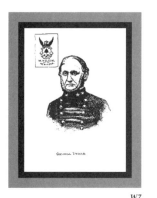

W7   W8

W7 CONFEDERATE GENERALS

Mortimer Nelson, New York, 1863. The face of each card in this deck features a portrait of an officer of the Army of the Confederacy or a high government official. It is interesting to note that a Northern company published this deck; perhaps at the time there were no Southern playing card manufacturers. The deck was likely made for sale in the South. Note the miniature card in each corner, an idea patented by Andrew Dougherty and used as early indices in his Triplicate decks a decade later.

W8 UNION GENERALS

M. Nelson, New York, 1863. This deck was issued at the same time and is the companion to W7. The Union Generals and Statesmen replace those of the Confederacy.

W9

W9 UNION GENERALS

National Picture Cards Co., New York, 1865. This is a completely different deck. The portraits are in a white oval that is centred on a grey card. The suits are identified by a red or black pip in the upper left and lower right corners of each card, making the cards reversible. The value of the card is shown in white within each pip. For example, a white 'K' within a Spade would designate the King of Spades. National Picture Card Co. was in reality M. Nelson. This deck was a publication of the 40 pictures used on the pip cards of W8 with allegorical courts similar in feeling to W5 and W6.

W10 CONFEDERATE DECK

Chas. Goodall, London, 1863. This deck is listed here as it was manufactured strictly for export to America. Once again the importance of playing cards to the public as well as to the fighting personnel is evidenced. American manufacturers were probably located in the North so the Southern States turned to England for their cards.

W10

W12 ARMY #303 *(not pictured)*
Russell, Morgan & Co., Cincinnati, Ohio, 1881. This was one of the earliest brands manufactured by the firm that was later to become the United States Playing Card Company. It featured a special Spade Ace and Joker both of which were dedicated to the Army (refer to US3 and US3a).

W13 NAVY #303 *(not pictured)*
Russell, Morgan & Co., Cincinnati, Ohio, 1881. This deck was published as a companion to W12 (refer US4 and US4a).

W14 ARMY & NAVY #303 *(not pictured)*
Russell & Morgan Co., 1885. This was a combination of W12 and W13 to make one brand honoring all of the armed services of that period (refer US5, US5a, etc.). Various editions of this deck were produced in later years by Russell & Morgan Ptg. Co. and USPC until about 1915.

W15

W15 ANHEUSER-BUSCH SPANISH AMERICAN WAR

Gray Lithographing Co., 1898. A most interesting war deck with two American officers on each Jack and one officer on the Kings. The Queens are 'Our Colonies'. Colonel Theodore Roosevelt shares the Jack of Spades with General Shafter while Admiral Dewey has sole possession of the King of Hearts. The background of each spot card has a picture of the Anheuser-Busch Brewery in light brown.

W11

W11 ARMY & NAVY

Andrew Dougherty & Co., New York, 1865. A most unique Civil War deck. Once again new suit signs were used with the blue suits depicting Monitors and Merrimacs. The Ace of Monitors reads "To commemorate the greatest event in Naval history, the substitution of iron for wood". The red suits are Zouaves and Drummer Boys.

W16

W18

W16 ANHEUSER-BUSCH, SPANISH AMERICAN WAR

USPC, 1900. This Spanish American War deck was issued at the conclusion of the war. The Jacks and the Kings both bear the portraits of two officers. This time Colonel Roosevelt shares the Jack of Clubs with General Wood. Admiral Dewey and General Miles are on the King of Hearts. Each Ace features a different brand of Anheuser-Busch beer. Finally, on each four spot there is an inset, sepia toned photo of a U.S. warship.

W18 DUTTON'S MILITARY CARDS

c1900. An unusual war deck featuring courts representing the Commander-in-Chief down to privates from several wars. The suits are the insignia of the Cavalry, Infantry, Artillery and Engineers.

W19 RUSSELL'S REGULARS *(not pictured)*

Willis W. Russell, Milltown, N.J., 1906. This was a standard deck of cards (refer RU5) in tribute to the men of the 'Regular Army'. Both the Ace of Spades and the Joker portray men in uniform.

W20 CRUISER #96 *(not pictured)*

Andrew Dougherty & Co., c1910. An early brand of this company's cards (refer AD34) gave tribute to the Navy.

W21

W21 ALLIED ARMIES

Montreal Litho, 1916. This deck was issued before the U.S. entry into World War I. The backs display flags of France, Belgium, Serbia, England, Japan, Russia and Italy. In this version spades represent Italy, hearts - Belgium, diamonds - England and clubs - Russia. The Kings and Queens were the actual monarchs of the respective countries and the Jacks were uniformed privates. The Ace of Spades features King Peter of Serbia, the Aces of Hearts and Diamonds have thumbnail sketches of President and Mme. Poincare of France. They were not placed on the court cards as they were not royalty.

There were actually six versions of the Allied Armies deck. To simplify matters, we will call all versions that have the Italians on the Spade courts, W21, and those with the Japanese, W22. This was not necessarily the order in which they were produced.

W21a ALLIED ARMIES *(not pictured)*

Montreal Litho, 1916. This back had six flags. President Poincare was on the Ace of Spades. King Peter of Serbia was on the Ace of Hearts and it had a blank Ace of Diamonds.

W17

W17 SPANISH AMERICAN WAR

NYCC, 1898. This deck further illustrates the need for our troops to be adequately supplied with playing cards. This is a standard deck with the box in the shape of a military knapsack. The 'blanket' roll at top of the box had a roll of small poker chips. It has a colorful patriotic (both factions!) back.

W21b ALLIED ARMIES *(not pictured)*
Montreal Litho, 1916. This issue had Madame Poincare added to the Ace of Diamonds.

W21c ALLIED ARMIES *(not pictured)*
Montreal Litho, 1917. The same issue as above except that borders were added to all court cards.

W22a

W22a ALLIED ARMIES
Montreal Litho, 1916. The first version of the same deck issued the same year, with the Japanese, not the Italians on the spade suit and President Poincare on the Ace of Spades. King Victor Emannuel III of Italy replaced Madam Poincare on the Ace of Diamonds (some versions have a plain Diamond Ace). This version had six flags on the back.

W22b ALLIED ARMIES *(not pictured)*
Montreal Litho, 1916 (listed originally as W22). The backs had seven flags as Italy had entered the war. The Japanese still have the spade suit, but King Victor Emannuel III of Italy is featured on the Ace of Diamonds. Peter of Serbia was now on the Ace of Spades and President Poincare on the Ace of Hearts.

W22c DOMINION RUBBER SYSTEM *(not pictured)*
Montreal Litho, c1915. This advertising deck for Dominion Rubber System (see CDN31) has courts reminiscent of W21 and W22 and is therefore listed here. It has a great advertising back and ads on the face of each card.

W22d

W22d ALLIED ARMIES
Montreal Lithographing Co., 1916. We have a copy of W22a with advertising on the back for Sunlight Soap that comes in a special Christmas box - clearly made to be sent as a Christmas present to soldiers overseas.

W23

W23 LIBERTY PLAYING CARDS
Liberty Playing Card Co., New York, 1915. They were designed as 'American Cards for American People'. They appeared to be issued to try to incite the public to press for U.S. involvement in World War I. The Joker shows Uncle Sam rolling up his sleeves with the title 'Invincible'. Bullets replace the aces, Commanders (two Admirals and two Generals) the kings, Nurses become the queens and Soldiers and Sailors, the jacks.

W24

W24 FREEDOM PLAYING CARDS
Freedom Playing Card Co., Portland, Oregon, 1917. In this unusual deck Uncle Sam replaces the kings, Liberty the queens and Infantry replace the jacks. The Joker shows a soldier and a sailor holding a donkey with the statement "No Kings or Queens for Me".

W25

W25 DEMOCRACY PLAYING CARDS
Democracy Playing Card Co., New York, 1918. It seems that each new deck issued for WWI was brought out by a new and different company, but it is probable that a larger manufacturer made the cards for these companies. The Democracy deck is loaded with slogans. "Make the world safe for democracy" and "Out with the Kings and Queens" to mention just a few. The kings show Soldiers, the queens are Nurses and the jacks, Sailors. The Aces are 'Our Aces of the Air'. The backs feature insets which relate to the different branches of the Armed Forces.

W26

W26 PLAYANLERN
Levis & Cook, Chicago, 1918. An unusual WWI deck which features a Soldier, Sailor and the Allied flags on the backs. The purpose of this deck was to teach French to our fighting men as they played cards. With the exception of the indices and the Ace of Spades every card was printed in light grey or pink. There were several phrases in English, translated to French, on each face.

W27

W27 MILITARY FORTUNE TELLERS
H.V. Loring, Chicago, 1918. This deck combined the fortune telling craze of the period with a salute to the Armed Services through the use of new suit signs. The blue suits were doves and bells and the red suits hearts and stars. Soldiers replaced the kings, Nurses replaced the queens, Sailors the jacks, and the aces were replaced by Aviatrixes. There were 56 cards in the deck as there were 1's in addition to aces.

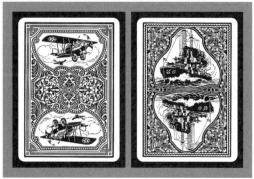

W28

W28 BICYCLE #808
USPC, 1918. The United States Playing Card Company in the spirit of the War, dedicated four special backs of the Nation's all time favorite cards to the Armed forces. A) Big Gun represented the Artillery, B) Dreadnaught, the Navy, C) Flying Ace, the Air Corps, and D) Invincible, the Tank Corps.

W29

W29 HISTORY OF THE GRAND ARMY OF THE REPUBLIC
Historical Playing Card Co., Indianapolis, c1920. A very interesting deck in which each card features the insignia and commanders from a different division of the GAR (1861 to 1865). The backs show a portrait of Abraham Lincoln who was President during the war.

W30

W30 MADEMOISELLE FROM ARMENTIERRE
Press of the Woolly Whale, 1933. One of the most interesting of the war decks. The owner of this publishing house, Melbert B. Cary Jr. was an ardent collector of playing cards and his extensive collection is now owned and on exhibition at Yale University. Press of the Woolly Whale also published his book entitled 'War Cards' in 1937. It was most fitting that he should also publish this war deck issued in commemoration of the signing of the Armistice and which honored the 103rd Field Artillery of which Mr. Cary was a member. The most interesting courts are the queens which represent the four Madamoiselles made famous by the popular WWI song ,'Parlez Vous'.

W31

W31 SUBMARINE DECK
Brown & Bigelow, 1941. This deck again emphasises the importance of playing cards to the average serviceman. The box states: "The Heart and Diamond suits on these cards have been specially printed for high visibility when players are wearing Red Adaptation Goggles used aboard U.S. Submarines."

W32

W32 AIRCRAFT SPOTTERS
USPC, 1942. This is the first in a series of three issues all by USPC. This wide deck with each face card having a silhouette of an aircraft used in WWII was designed in the hope that the cards would educate the public in aircraft recognition while they were enjoying their favorite games. This version shows two views on each card.

W33

W33 AIRCRAFT SPOTTERS
USPC, 1942. The second version of W32 with a view of the planes from the ground, the view ground observers would be most likely to see.

W34

W34 AIRCRAFT SPOTTERS
USPC for the Coca-Cola Company, 1942. This was a narrow version of these decks with three views on each card. The aces, jokers and backs (there is also a Nurse back for this deck) advertised Coca-Cola.

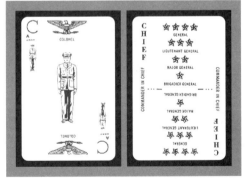

W35

W35 ANMA
Anma Card Co., Tulsa, Oklahoma, 1941. The name of the deck and the company was created using the initials of the four branches of the service, Army, Navy, Marines and Air Corps. The red suits are Army and Air Corps and the blue suits are Navy and Marines. The courts feature aces as Colonels, kings as Lt. Colonels, queens as Nurses and jacks as Majors. The deck is unusual as each spot card is given a rank from the 10 (Captain) to the 2 (Private 1st Class). The Commander-in-Chief is used in place of the Joker.

W35a

W35a ANMA
Anma Card Co., Tulsa, Oklahoma, 1941. There was a second edition published the same year. The indices are a different style and this deck has four color indices. The Army's are red; the Navy, black; the Airforce, green; and the Marines, blue. The box emphasizes that any game could be played with this deck.

W36

W36 VICTORY PLAYING CARDS
ARRCO, Chicago, 1945. This deck was issued right after VE Day (Victory in Europe). The kings represent Uncle Sam, the queens the Statue of Liberty and the jacks feature a Soldier and Sailor on each card. The Joker is a caricature of Hitler and Mussolini.

W37

W37 FORBIDDEN CITY, PEKING & CHINESE VIEWS (BOXER REBELLION)
Grimes-Stassforth Stationery Co., Los Angeles, 1910. This deck features views of China but it was made in and sold only in the USA. It is classified as a war deck as the faces feature oval photo scenes of China and events from the Boxer Rebellion. The backs have a reversible design showing yellow pennants with a blue dragon on a background of peacock feathers. This deck could be called a souvenir deck of the Boxer Rebellion.

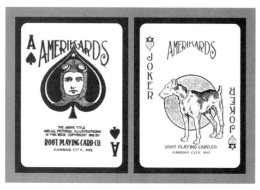

W38

W38 AMERIKARDS
Root Playing Card Co., Kansas City, Mo., 1919. One of the last WWI decks and one of the most interesting. Each ace features a picture of an Aviator. The kings are Soldiers, the queens, Nurses and the jacks, Sailors. The Joker is a member of the Canine Corps.

W41

W41 GAME OF PEACE
Peace Playing Card Co., Lansing, Michigan, 1912. The company advertised that one could play "six games of peace and all standard games" with this deck. The suits were Germany, France, England and the United States and the deck was published by Homer L. Boyle, author of 'History of Peace'.

W39

W39 THE KENT CIGARETTE DECK
The backs and case look like a pack of Kent Cigarettes. All of the faces show members of the Armed Services. The Privates are aces and the kings are Generals. The spade suit represents the Navy, hearts the Air Force, diamonds the Marines and clubs the Army. The queens represent the Women's Auxiliary of the particular service. The two jokers feature the Cook and Re-enlistment Sergeant.

W42

W42 WAR SCENES
Mortimer Nelson Co., 1863. A rarely seen satirical civil war deck was produced by Nelson as one of a series of decks with scenes. The cards mix romantic, comic and tragic war observations. While the title of the Ace of Spades is 'End of the War', evidence is strong that it was issued in 1863, before the end! Other known decks from the series are discussed under N48.

W40

W40 MILITAC
Parker Bros., Salem, Mass., 1916. The four suits in this deck are the four branches of the Army. Artillery in red, Cavalry in yellow, Engineers in green and Infantry in blue. The ranks are from Private (1) to Lieutenant Gen. (12). There is a booklet with rules for special games that closely resemble bridge. There is also an identical deck, Tactics by the National Military Tactics Co., NY, 1916. As it was probably manufactured by Parker Bros. who held the copyright, our number remains the same.

W43

W43 TEDDY CARDS
Chicago Playing Card Co., Chicago, 1898. A standard deck (refer L24) with backs showing a portrait of Teddy Roosevelt in the upper right hand corner. The rest of the back has a scene from the Spanish American War with Roosevelt on a horse.

W44

W44 AVIATION PLAYING CARDS
1972. A later deck put out as a tribute to aviation and its personalities, many of whom earned their reputation from war exploits. It was published by the Insurance Company of North America.

W47

W47 WILLIAM AIKEN WALKER CONFEDERATE CARDS
c1864. A hand painted deck where each card is an original painting by the famous southern artist. Each card is beautifully executed and the painting is bright and colorful. Two copies of this deck are known.

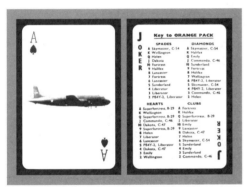

W45

W45 RECOGNITION CARDS
1942. Another recognition deck known also as the Orange deck and perhaps made in Canada. The Joker matches each card with the name of its plane.

W48 VICTORS *(not pictured)*
USPC, 1919. This standard deck (and the two that follow) was issued in honor of and named for the Allied Victory in WWI (refer US31). It was sold in the U.S. and also in the new London offices of USPC (note the addresses listed on the Ace of Spades).

W49 CANTEEN #515 *(not pictured)*
USPC, 1898. This brand was created for the Spanish-American War. It was a non-enameled, inexpensive deck (refer US29), made to be sold to the service men at the Armed Forces canteens.

W46

W46 HUNT AMERICAN MANUFACTURE
c1815. This deck, likely commissioned by an English maker for sale in the United States, has a wonderful 'Decatur Victory' Ace of Spades. The rest of the deck is standard.

W50 PICKET #515 *(not pictured)*
USPC - RM Fact., 1914. Another deck for soldiers was introduced in 1914 using the same style number as Canteen. The name was changed to Picket and the Joker portrayed a 'Doughboy' of World War I (refer US30).

A number of manufacturers in the 1860s and 1870s produced standard decks with backs that had war interest or commemorated events from the war. One of the first U.S. issues with pictorial instead of pattern backs manufactured by John J. Levy, New York, was the naval battle pictured here from the War of 1812. The next two pictures are of interest. The interesting back with flags and small portraits is the Three of Diamonds upon which is written "A Momento of the Battle of Somerset, Kentucky Taken from the Jacket of a Dead Rebel March 30 1863". The very colorful flag back surrounded by a patriotic banner is one of the design used by Dougherty for its illuminated deck. The last back is from a very interesting deck patented in 1869 on which each face is a different ad for a Boston area business (see A49). The rest of the backs shown are from standard decks.

POLITICAL AND PATRIOTIC CARDS

Political and patriotic themes were prevalent in playing cards long before the United States was a country. As far back as the 16th century playing cards have honored the rulers of nations and heraldic cards have paid tribute to the politicians of their day.

It is therefore quite surprising that so few American decks fall into this category. We could consider war cards as Political and Patriotic but we believe that the category in which they have been placed is the most appropriate.

Since the early 19th century, American playing cards have been used as advertisements for political campaigns. We are not concerned with those decks that merely used the backs as campaign media and we will ignore those issues with a portrait decorating the back. Others, which may have a political logo on the Joker or the Ace of Spades, likewise do not arouse the public interest enough to list here.

Only decks with sufficient face interest are included in this section. The face interest may be shown on the court cards, on a particularly unusual Ace of Spades, or perhaps in a theme that will extend to every card in the deck. Oddly enough, in recent years we have seen the largest number of political decks issued. Perhaps this is due to the fact that it is now possible to satirize and ridicule our political leaders, in a sense of good spirit.

P1

P1 GEORGE AND MARTHA WASHINGTON

Samuel Hart & Co., New York/Philadelphia, 1866 to 1871. These decks had beautifully designed Aces of Spades (refer NY32, NY33, NY34, and NY47a). The original issue in 1866 (NY32) featured the portraits in full color. In a second edition, issued at approximately the same time, a black and white Ace of Spades was used to reduce the cost of production. However this copy of the original Ace did not look distinct in black and white so a new design (NY33) was created producing a much younger looking George and Martha. In 1871, to inaugurate the newly formed New York Consolidated Card Company, of which Samuel Hart was a key member, the next issue (NY34) was fully colored and produced in a deck that included illuminated courts and pips (like NY31) on the spot cards. All of these issues had two way courts and no indices.

P2

P2 NATIONAL CARDS

1883. This is an interesting deck that was a later copy of the ideas originated by B.W. Hitchcock for the American Card Co. of New York in 1863. He obtained the copyright for this version in 1883. You may recall the decks in the War chapter (W5 and W6) which introduced Eagles, Flags, Stars and Shields as suit signs as an attempt to change the traditional royalty courts. The deck is interesting and political in nature. The kings represent George Washington, Andrew Jackson, Abraham Lincoln and Chester Arthur. The exact date of the issue is unknown, but the appearance of President Arthur would indicate he was President at the time. This deck is often referred to as 'The Bad Joker'.

P3

P3 COMIC POLITICAL PLAYING CARDS

A.H. Caffee, New York, 1888. This favorite of collectors was possibly designed for the presidential campaign of Grover Cleveland in his effort to regain the presidency from Benjamin Harrison in 1892. However many collectors believe both it and P18 were from the 1888 campaign. All of the court cards are political figures on the national scene. The King of Hearts features Cleveland and the King of Spades, Harrison. A caricature of General Butler graces the Ace of Spades. The Joker shows a boxing scene with Cleveland overcoming Harrison. An extra card was issued with the deck identifying the different court cards. A second deck, unknown at the time of the original Encyclopedia section, promotes the Harrison side and is listed as P18. The backs of both decks are the same except that the Cleveland deck is orange and the Harrison deck blue. Interestingly, the pips for both decks are of the Triplicate type.

P4

P4 PRESIDENT SUSPENDER
C.A. Edgarton Co., Shirley, Mass., 1904. This was an advertising deck for President Suspenders. Each king and jack features two different presidents and each queen, two first ladies. George Washington is on the Joker. The backs advertise the President Suspender Company.

P7

P7 BANNISTER BABIES
Brown & Bigelow, 1960. Political titles are subtitled under babies with appropriate facial expressions.

P5

P5 ROYAL REVELERS
Brown & Bigelow, 1932. This deck was issued to aid in the fight to repeal the Prohibition Amendment to the Constitution. The courts hold glasses and definitely appear to be inebriated. The Joker is Mr. Bluenose, which was a pseudonym for the Censor.

P8

P8 KENNEDY KARDS
Humor House, 1963. An interesting and amusing deck in which all of the court cards are members of the Kennedy family or members of his administration. This deck was never released for public sale. The tragic assassination took place just before the scheduled release and the publishers thought it would be in poor taste. They did find their way to the public eventually.

P6

P6 GEORGE WASHINGTON PLAYING CARDS
American Playing Card Corporation (NYCC), Portland, Maine, 1934. A colorful deck with the Kings depicting different sketches of George Washington and the Queens, his wife Martha. The Jacks feature laborers in different occupations.

P9

P9 THE TEXAS WHITEHOUSE
E. & S. Co., Austin, Texas 1966. This deck was devoted to historical facts about Texas and every card was filled with information. The backs featured 'The Texas White House', home of President Lyndon B. Johnson and his initials, LBJ.

P10 GOLDWATER FOR PRESIDENT (*not pictured*)
1966. A unique political campaign deck. Every one of the court cards features the face of Barry Goldwater superimposed on the face of normal courts. The backs (see A8) come in two colors, white and black with the chemical symbols AuH2O (for 'gold' and 'water') in the center.

P14

P11 QUEEN HIGH EQUALITY DECK
Emjay Co., 1971. A women's liberation deck which states, "Why play with a male-dominated deck?" In this Equality deck the Jacks are no longer princes, rather Jackie, a princess. Of course the Queen is high!

P11

P14 EXECUTIVE DECK
Mercer & Smith, printed in Japan, 1973. Another deck made with caricatures of Nixon and his political family. The Kings are Nixon, the Queens, his wife and the Jacks are various other politicos.

P12

P12 POLITICARDS
Politicards Corp., Los Angeles, 1971. Every card in the deck is a caricature of a well-known figure on the political scene. The red suits are devoted to the Democrats and the black suits to the Republicans. The spade suit is reserved for President Nixon and his administration.

P15

P15 VOTES FOR WOMEN
published by the National Woman Suffrage Publishing Co., 171 Madison Avenue, New York City, c1910. The Ace of Spades, the Joker, the backs and the box all promote 'Votes for Women'.

P13

P13 THE PRESIDENT'S DECK
Alfabet Co. (printed by USPC), Portland, Oregon, 1972. Richard Nixon is featured on the Kings, Mrs. Nixon on the Queens and Spiro Agnew on the Jacks. The Jokers are Wallace and Humphrey.

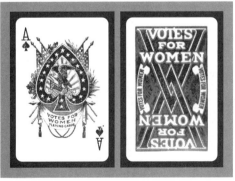

P15a

P15a VOTES FOR WOMEN II
A second deck, published for the same cause, was produced by American Playing Card Co. of Kalamazoo. Both these decks are scarce and highly sought after by collectors of Suffrage Movement memorabilia.

P16

P18

P16 '49'

Liberty Games Co., Inc., Versailles, Mo., 1920. A very unusual political deck named for the 48 states plus the District of Columbia. The rules for the game are supplied, but the deck is designed so that regular card games can be played as well. The 'K' in place of the king is the first letter of 'Known as the chief chosen person of the Party' and the presidential candidate is pictured. The 'Q' stands for Quality, second place, so the vice-presidential candidate is shown. The 'J' stands for Jester and the Motto and the symbol of the parties are shown. The aces feature the party platforms. The major parties, Democrat and Republican, are the blue suits while the secondary parties, Socialist and Farmer Labor, are the red suits. The female Joker was supposed to appeal to woman voters.

P18 COMIC POLITICAL PLAYING CARDS

A.H. Caffee, New York, 1888. The other deck for the 1888 campaign that is like P3 but for the Harrison side. The red Kings are Harrison and Morton (the good guys) and the black Kings are caricatures of the bad guys, Cleveland and Thurman. It had the same Ace of Spades as P3 and a similar Joker, but with Harrison knocking out Cleveland. An extra card was issued with the deck identifying all of the different court cards. The backs of both P3 and this deck are the same except that the Cleveland deck is orange and the Harrison deck blue.

P17

P19

P17 CLEVELAND CAMPAIGN

1892. This deck was issued during the 1892 campaign. The deck is serious, not comic and was issued to appease those party members who objected to the comic aspect of P3. The Ace of Spades says 'Campaign' and the court cards feature Cleveland, Mrs. Cleveland and Thurman, who was the candidate for Vice President.

P19 TIME MAGAZINE

1978. This reproduction of a very early American deck has indices added for modern convenience. The backs copy an invitation to an Independence Day Ball. The cards are modeled after an imported Hardy deck using an Amos Whitney Ace of Spades.

P17a CLEVELAND CAMPAIGN (not pictured)

1980. U.S. Games Systems issued a reprint of the Cleveland Campaign deck in 1980.

P20

P20a

P20 POLITICAL EUCHRE

L. Lum Smith, c1880. There are two versions of this deck known which were likely issued in 1883 for the Cleveland and Blaine presidential campaign of 1883-84. This version is 'patent pending' while P20a is 'patented'. On this deck the jacks are not named but represent the presidential and vice-presidential candidates. The Uncle Sam Joker is replaced on the second version by General Butler. The cards in this deck are of a higher quality and were perhaps made by Dougherty. We have pictured a fascinating advertisement for this version.

P20a POLITICAL EUCHRE

L. Lum Smith, c1883. This version with General Butler as the Joker was quite controversial and the General raised objections in a lawsuit about his treatment on this card. The cards in both decks are fascinating for the history they provide about the distribution of electoral votes, state populations and politics, etc. In this deck the four Jacks, the highest cards in Euchre, are the candidates for president and vice-president, Cleveland and Hendricks (Democrats) and Blaine and Logan (Republicans).

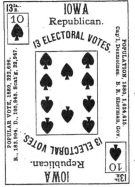

P21

P21 SOCIALIST PLAYING CARDS

Charles H. Kerr & Co., Chicago, 1908. An ad for this deck describes "a full pack of 53 playing cards ...... the 12 picture cards are original caricatures - the Kings on the trusts, the Queens on the capitalist virtues, and the Jacks on the principal Guardians of the Existing Orders. Most of the other cards in the pack stand for various types of workingmen and women, and an appropriate rhyme is printed on each".

P22

P22 ROYAL PLAYING CARDS

(Formerly SX4), NYCC, c1894. This deck was originally thought to have been issued for the Paris Exposition in 1878 and therefore originally listed in the Exposition and World's Fair chapter. It has subsequently been demonstrated (refer Clear The Decks, Vol. VII, No.2) that it was published in the mid-1890s as a deck commemorating the Triple Alliance of Germany, Austria and Russia (subsequently replaced by Italy) and the suits are dedicated to these four powers. The clubs show personages from Germany; the diamonds, Austrians; the hearts, Italians; and the spades, Russians (with the exception that the queens are reversed on the heart and spade suits). It is one of the most beautiful American decks ever manufactured with its fine engraving, many colors and liberal use of gold.

P23

P23 CHAS. P. HART

c1920. The deck has four suits – 'R' for Republics, 'D' for Democrats, 'S' for Socialists and 'P' for ...? We are unsure about the name of the last suit and have pictured the Originator of 'P', Theo. Roosevelt, for readers to determine. Also pictured are the 10 of Democrats and the Joker, a card honoring the Suffrage Movement. The backs show an autographed portrait of Chas. P. Hart.

ENTERTAINMENT

From 1890 to 1915 is considered to be the heyday of wide souvenir playing card manufacturing and various types of these cards are covered in Chapters 24 to 27. Another form of souvenir card relates to entertainment activities of different kinds and this chapter discusses these decks which are, in some cases, very valuable and difficult to find.

It is impossible to determine just how many events had special souvenir cards printed to distribute or to sell. Most of these were limited editions and may have totally disappeared years ago.

The most common of the entertainment decks are the sport, stage and movie decks that depict the stars or sport figures of the era. Again we face a crossover of interest as these decks are actively sought after by movie,

sport and nostalgia collectors, as well as those who collect playing cards.

In later years many narrow Entertainment decks were issued. Movie stars, pop and country music singers, television programs, sporting events, and countless other fields of entertainment are all readily available to the collector. These decks are collectible but generally not very valuable and it would be a huge job to catalogue them all. Likely this field will be one of the most sought after categories in the future.

Generally the decks considered in this section are wide issues or those published prior to 1935, although we show a quite a few examples of narrow, later decks. Almost all have face interest but we have also included a small number with very specialized back designs.

SE1

SE1 BABES IN THE WOODS
George H. Walker & Co., Boston, 1893. This beautiful deck was issued as a souvenir of the 75th performance of 'Babes in the Woods' on January 10th, 1893 at the Boston Theatre. The pip cards are standard but all the courts and the Joker portray characters from the play and are printed in color. This deck was given as a momento to all the patrons who attended this performance. As it was a very limited edition, it is rare.

SE2

SE3

SE2 THE CRADDOCK SOAP STAGE SOUVENIR DECK
USPC, 1895. This deck was published as an advertising deck for Craddock's Medicated Blue Soap. The faces feature portrait photos of 53 Stage Stars of that era. The Queen of Spades is Julia Marlowe and the Joker, De Wolf Hopper.

SE3 STAGE #65
USPC, 1896. All of the aces and the court cards feature four pictures of Stage Stars. The courts are set in circles and reversed with a crown at both ends. The Joker is Marshall P. Wilder. This deck is very similar to the Monarch Bicycle advertising deck (A12).

SE4

SE7

SE4 STAGE #65

USPC, 1896. This is a similar edition of the same deck. The only difference is that, in this issue, all of the jacks, queens and kings have their denominations indicated within their pip. As both of these decks were made until 1900 there may be a change or two in the photos on either deck.

SE7 MOVIE SOUVENIR PLAYING CARDS

M.J. Moriarty, 1916. This deck features the '53 Most Prominent Stars in Filmdom'. The fact is that there were many different stars shown on different decks leading one to believe that there were many more than 53 prominent stars! The deck was published for several years and an alternative theory is that fame was quite fleeting in early days of 'Filmdom'. The backs feature a painting of a chariot race and the Joker in every deck is Charlie Chaplin. An interesting piece of trivia is that Mr. Moriarty was the purchasing agent for USPC who printed the cards for sale and distribution by him.

SE5

SE5 STAGE PLAYING CARDS #65X

USPC, 1908. This beautiful deck features one star on each of the 53 cards. The stars are centered in half tone oval photos framed in gold. They are bordered along the sides with different floral arrangements for each suit. The Queen of Hearts features Julia Marlowe and the Joker is Marshall Wilder.

SE8

SE8 BASEBALL PLAYING CARDS

The Baseball Card Co., New York, 1888. There were quite a few baseball decks issued in this very early period, however SE8 is the only one known to have miniature playing card symbols on the faces. The deck consisted of 72 cards portraying the nine starting players of the eight American Baseball Teams of 1888. The cards go from six to ace representing the nine different positions of the players (e.g. aces are catchers, kings are pitchers, etc.). This deck is extremely rare and even the individual cards command a premium price.

SE6

SE6 JEFFRIES CHAMPIONSHIP FIGHT SOUVENIR CARDS

W.P. Jeffries, Los Angeles, 1909. The backs feature a portrait of James J. Jeffries in a derby hat. The faces have oval photo scenes of famous bouts and action pictures of the top prizefighters of the day.

SE9

SE12

SE9 MOLLY O
USPC, 1921. The back features sepia toned photos of the star and is titled 'Molly O'. The faces have rectangular photo scenes from this silent movie hit. The deck was issued to promote the movie and was probably given away.

SE12 BASEBALLIZED PLAYING CARDS
Patented Oct. 13, 1925. An interesting deck of cards where every card shows a different 'play' from the game. This deck can be used for playing a simulated game of baseball as well as for playing regular card games because of the retention of all suit and rank signs.

SE10

SE13

SE10 10th OLYMPIAD PLAYING CARDS
P.G. Wenger, 1932. The backs feature an Olympic participant carrying a garland. Two different film stars are shown on each court card. Flags of the different participating nations are on the spot cards and the views of the stadium are pictured on the aces. The pack was issued to promote and raise funds for the Olympic Games held in Los Angeles in 1932. The Joker is Joe E. Brown.

SE13 OFFICIAL FILMS, Inc.
1957. A special printing for this film company which produced 'Four Star Playhouse', a television series. Each card bears a likeness of a movie or television star and their featured players, e.g. Ida Lupino and Dick Powell, were used on more than one card.

SE11

SE14

SE11 STAGE STARS of 1882
The Playing Card Novelty Co., Washington, DC. This deck was issued as an advertising deck for Duke's Cross Cut Cigarettes, a product of W. Duke, Sons & Co., Durham, N.C. It is interesting to note that this deck was issued during the time of prolific tobacco advertising. Many tobacco firms were including, in their product, insert cards with playing card symbols and beauties of the day. This deck was the only one which included the names of the stars.

SE14 UNIVERSAL STARS PLAYING CARDS
designed by Alexander Paal for Universal Studios, 1941. Each card bears a reversible photo of a different studio star. The Joker has an advertisement for their movie picture 'Nice Girl?' starring Deanna Durbin.

SE15

SE15 HOLLYWOOD FORTUNE TELLING CARDS

c1935. A miniature deck which is quite unusual as the face cards are standard, but each back has a photo of a different movie star. There is no maker name on the cards and there does not seem to be any reason for the term 'Fortune Telling' cards.

SE18

SE18 BASEBALL BACKS

National PCC, c1900. The two cards pictured are similar examples honoring baseball. They feature bats, balls, caps and other equipment.

SE16

SE16 BASEBALL BACKS

Willis Russell, Milltown, N.J., 1906. Willis Russell issued three decks with different photos of baseball players on the backs of their 'Rustlers' brand. We cannot determine if these photos represented particular players, or just players fielding different positions. The Joker also represented a player.

SE19   SE19a

SE19 HOME RUN

Pyramid Playing Card Co., Brooklyn, NY, c1930. This company made two standard decks with baseball themes. This is their wide variety.

SE19a UMPIRE

Pyramid Playing Card Co., Brooklyn, c1935. This is the narrow variety.

SE17

SE20

SE17 BASEBALL BACKS

N.Y. Consolidated, c1900. A series of backs featuring the national pastime in which scenes of games and baseball equipment were the feature of the backs of the cards.

SE20 BLACK CROOK PLAYING CARDS

1893. SE1 'Babes in the Woods' was issued as a souvenir of its 75th performance, to patrons of the Boston Theatre in 1893 by the owner, Eugene Tompkins. He repeated his promotion with 'The Black Crook' deck, which was issued on November 28th, 1893, to honor the 100th performance of this successful play.

SE21

SE21 BROWN DERBY PLAYING CARDS

Brown and Bigelow, 1950. This deck was issued for The Brown Derby restaurant in Hollywood. Each card features a caricature of a noted entertainer and was drawn by several different artists. The proceeds were donated to the City of Hope hospital, dedicated to doing research on childhood diseases.

SE22

SE22 W.C. FIELDS COMMEMORATIVE DECK

J.L. BROWN, 1971. This deck features W.C. Fields with scenes and quotations from his films on the courts and aces. He is also shown on the back and Joker.

The cards under the SE23 listing categories, Western stars, baseball a etc., and were sold in postcard size sh pictured four playing card photos wit backgrounds. Research by John Lafl there are at least nine series, some in different sizes. They are:

1. Cowboys in a circular vignette with a postcard back

2. Cowboys with vignette contained in the pip shape and with a plain back

3. Cowboys, plain back and a gray or yellow background

4. Cowboys with various background colors issued four up on postcards

5. Leading men with various background colors and plain backs

6. Mostly baseball players (some other celebrities) with cream background

7. Cowboys with various backgrounds issued 5 inches by 2 inches on postcards

8. Similar to #6 but in two colors

9. Similar to #6 with screen stars

SE23

SE23 ARCADE CARDS

c1929. A few representative cards are shown.

SE24

SE24 THE MAN FROM UNCLE
Ed-U-Cards, 1965. Copyright by Metro-Goldwyn-Mayer, the producers of the TV series. The courts feature the stars of the show and all other cards have action scenes adapted from the shows.

SE27

SE27 POP MUSIC
Heather Publications, Denver, 1966. Each card has a photo of a famous personality of popular music from Sinatra to Elvis, to the Beatles.

SE25

SE25 THE GREEN HORNET
Ed-U-Cards, 1966. A similar deck for another popular TV series.

SE28

SE28 COUNTRY MUSIC
Heather Publications, 1967. This deck was released in two editions, the second in 1970 by Brown & Bigelow for Heather.

SE26

SE26 LAUGH-IN
Stancraft, 1969. Another popular TV comedy show with the faces of the stars superimposed on the court cards. The pip cards are covered with one liners from the show.

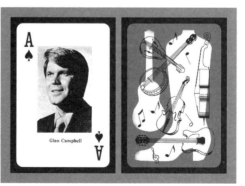

SE28a

SE28a COUNTRY MUSIC
Heather Publications, 1967. A second edition.

SE29

SE29 THE MONKEES
Ed-U-Cards, 1966 for Raybent Productions. The courts are standard but the pip cards feature action scenes and the Joker shows the group.

SE32

SE32 WESTERN STARS
c1960. Postcard size deck which comes with plain as well as postcard backs that can be written on.

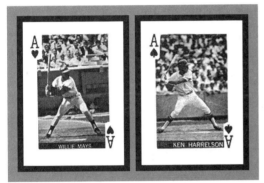

SE30

SE30 MINATURE BASEBALL STARS
c1965. A thin, cheaply made photographic deck with baseball players on each card.

SE33

SE33 FLICKERS
Creative Impressions, Los Angeles, 1974. This deck features scenes from Metro-Goldwyn-Mayer films during 1929-1959. The scenes are repeated in each suit.

SE31

SE31 FAMOUS PERSONALITIES
c1965. As above, only these cards feature famous people both alive and dead.

SE34

SE34 NFL GREATS
Stancraft, 1970. A different all-time great player from the NFL is featured on each card and the aces show the NFL insignia.

SE35

SE35 MILWAUKEE BUCKS
Each card features a player from this basketball team and the stars are repeated in each suit. The Aces show the team name and a sketch.

SE36

SE36 SHIRLEY TEMPLE
USPC, c1935. USPC made several special decks to profit from the popularity of this young star.

SE37

SE37 MOVIE STARS
Alfred F. Howe, New York, 1920. A rare deck with a different star on each card with "peeps at your favorites on and behind the screen – Interesting Facts Not Generally Known About 53 of the Foremost Film Players and Sidelights on Many Others".

TAROT AND FORTUNE TELLING CARDS

These two categories of playing cards, while often considered together, are not the same and need be listed and described separately. There have not been a large number of either produced in America and they have therefore been included in the same chapter.

The earliest known playing cards in Europe, some dating back as early as 1440, were tarot cards. They were made for playing the game of Tarot and not for telling fortunes. A Tarot pack consists of 78 cards based on the Italian suit symbols of Coins, Cups, Swords and Batons. Each suit consists of 14 cards, 10 spot cards (ace to ten) and four court cards (king, queen, knight and knave). In addition, there are 22 trump cards, often called atouts, which rank higher than the other 56 cards. The early trump cards portray fanciful figures - Magician, Devil, Hanging Man, Death, Wheel of Fortune, Lovers, etc. Historians theorize that playing cards were originally introduced into Europe from the East and legend credits the introduction to bands of gypsies and returning crusaders. Perhaps due to the strange and mystic symbolism that appears on the atouts, people used them to predict the future and tell fortunes.

Closely related to tarot cards are the 54 card tarock decks. The game of Tarock is a popular game in Austria, parts of Germany and in Switzerland. The pack retains the 22 trump cards but the mystic characters are replaced by themes. Tarock cards have atouts that show animals, ballet dancers, war scenes, cities, comic themes, etc. It is interesting that tarock decks usually have the French suits of Spades, Hearts, Diamonds and Clubs. The 22 atouts, the king, queen, knight and knave of each suit as well as the ace, two, three and four of the red suits plus the seven, eight, nine and ten of the black suits make up a complete deck. Tarock decks were rarely used for fortune telling.

There are only a few early tarot decks that originated in the United States. In the past 20 or so years, however, tarot cards in the U.S. have enjoyed increasing popularity. This is due in great part to the efforts of Stuart Kaplan of U.S. Games who has written numerous books on the subject and is responsible for the reprinting of many older decks. He has also inspired the designing of numerous new ones. Those interested in tarot and its decks should certainly acquire *The Encyclopedia of Tarot* by Kaplan (Volume I published in 1978) and his subsequent volumes.

Although tarots date back to the 15th Century, the earliest known fortune telling cards are from the late 17th century. Their popularity did not peak, however, until the publication of the Le Normand Fortune Telling cards. Mme. Le Normand was a member of Napoleon's Court and as his personal fortuneteller, she was often consulted about crucial decisions of state. Shortly after this time every European cardmaker of note published their version of Mme. Le Normand's cards. The demand for the Le Normand cards lasted throughout the 19th and into the 20th centuries, and versions are still being published today.

The earliest American fortune telling deck dates from about 1820. Le Normand decks seemed to be as popular here as they were in Europe, and in addition, there were many unusual decks that featured new ideas. For example, the United States Playing Card Co. issued a partial transformation deck for fortune telling (Ye Witches) in 1896.

Following Gene Hochman's example, only fortune telling decks with suit signs will be included here and decks such as The Teuila Fortune Telling deck, first issued in 1899 and reissued in 1923 by U.S. Playing Card Co. will be excluded. A few new listings have been added since publication of the original Encyclopedia. Other 'take-offs' of the Le Normand decks that have originated since the 1930s have been left for a more comprehensive work.

In 1918, the De Laurence Company of Chicago issued the De Laurence Tarots. The designs were so identical to

TA1

TA1 RIDER-WAITE TAROT
Designed by Pamela Colman Smith, an American, in 1910. This deck was commissioned, and supervised by Arthur E. Waite, of London. Although this deck appears to have been first published in 1911 in London, some collectors believe that the identical deck was issued simultaneously in New York. It is unfortunate that no marks of identification can be found on the cards, either the originals or on the reprints issued over many years. Only those that are still in original cases can determine which edition they might be from.

the Rider-Waite decks that they might have been made from the same plates. Only a sharp difference in the color treatment distinguishes them, as they were cheaply printed in two colors only, yellow and black or red and black. There were many later reprints made in the United States and elsewhere, but they can usually be spotted at once by the thickness of the cards. An original Rider-Waite, measures 38mm when the 78 cards are stacked (the size of the original deck was 71x121 mm) while most reprints measure 23mm. American reprints have been published in the 1960s and 1970s by University Books, NY, Albano Productions, CA, and U. S. Games Systems, CT.

TA2 BUILDERS OF THE ADYTUM TAROT
1922. This is a black and white deck, similar in many ways to Rider-Waite and designed by Jessie Burns Parke, specifically for the Builders of the Adytum. The cards are 64x107 mm. A reissue of this deck by the same people was published c1970.

TA3 REVISED NEW ART TAROT CARDS
J. Augustus Knapp, 1929, Los Angeles. A beautiful deck done in four colors and although it retains the standard atout themes, the designs were completely different than any of the predecessors. A booklet was enclosed with each deck entitled 'An Essay on the Book of Thoth' by Manly P. Hall.

TA4 GREEN SPADE TAROT
August Petryl & Son, Chicago, 1922. This is a beautiful 74 card Tarot deck which includes the 52 cards from NR5 plus an additional 22 atouts, all portraying some phase of the life of American Indians. To quote from an instruction pamphlet included with the deck: "Tarok is the pioneer of all card games, now simplified and Americanized. Skill, not luck, makes for victory. Every minute of the game holds your undivided attention. Indian auction, a new, well balanced game in which a weak hand has an equal chance with a powerful hand. Fortune telling, simplified, yet thorough, and arranged to reveal the Past, Present, and Future in a weird and mysterious manner". There is also a 54 card Tarok deck with an extra court, a mounted Frontiersman, in each suit and a 78 card Tarot deck using the 74 card deck and adding the four mounted Frontiersmen.

This Encyclopedia is mainly restricted to coverage of decks from before WWII and the many tarot decks that have been issued in America in the last few decades are not included, although we have retained the listings of the two below from Part VI of the original.

TA5 NEW TAROT FOR THE AQUARIAN AGE
Western Star Press, Kentfield, California, c1968. This Tarot is also known as the Book of T. It has colorful and unusual atouts and the suit signs are Blades, Serpents, Stones and Pears. It is a full 78 card deck. The atouts are not numbered and the instructions are needed to discern the order.

TA6

FT3

TA6 NEW TAROT
*Hurley and Horler, Sausalito, California, 1974.
Unusual modern designs in black and white with
Circles, Wands, Cups and Swords in an odd
transformation style. The atouts are standard but very
modern. It is a complete 78 card deck with an
accompanying booklet which describes each card.*

FT3 MADAME LE NORMAND'S GIPSY FORTUNE TELLING CARDS
*Wehman Bros., NY, c1900. These cards date somewhat later than FT2, but have the
identical card designs. Phillipe again signs the instructions. The numbers and cards
are in red and the fortune telling designs and borders are in black.*

Turning to early Fortune Telling decks, we find they
were produced in the United States starting about 1820
and continuing through the 20th century.

FT1

FT4

FT1 AMERICAN FORTUNE TELLING CARDS
*Turner & Fisher, c1820. This deck is believed to be the earliest fortune
telling deck on the American scene, but the date cannot be determined
accurately. Gene Hochman had a copy in black and white dated 1862
and a delicately colored version that appeared much older. The same
plates were used for reprints throughout the years, the latest being
issued by Wehman Bros., New York, in 1948.*

FT4 MADAM LE NORMAND'S FORTUNE TELLING CARDS
*The Temple of Revelation, Gloversville, NY, c1900. This was possibly
one of the hundreds of smaller fortune telling emporiums exploiting
the 'believers'. The playing card inserts were more modern and the
numbers changed, although the same fortune telling symbols are used.*

**FT2 MADAME LE NORMAND
CARDS**
*McLoughlin Bros.,NY, c1880. The cards
were simple and printed in black and
white. They measured 58x83 mm. The
box was elaborately colored. The
instructions enclosed were signed by
Phillipe, an heir of Mlle. Le Normand.
Many American versions of this deck
have been published in the United States
between 1860 and this writing.*

FT2

**FT5 MADAM LE NORMAND
FORTUNE TELLING CARDS**
*Anonymous, NY, c1880. This is a
beautifully hand colored deck which came
packaged in a paper wrapper marked
Madam Le Normands, New York. The
cards and symbols follow the usual form
and the size of the cards is 60x93 mm.*

FT5

FT6

FT6 MLLE. LE NORMAND'S ONLY SURE AND COMPLETE FORTUNE TELLING CARDS

A. Nielen & Co., Cincinnati, c1900. This deck had quite a variance in layout and design. It utilized many of the same symbols, but others were replaced. It measured 53x92 mm.

FT7

FT7 MADAM MORROW'S FORTUNE TELLING CARDS

McLoughlin Bros, NY, 1886. This deck is almost identical to the Le Normand cards, but in black and red on a pink striped background.

FT8

FT8 COMIC PLAYING CARDS OF FORTUNE

E. Gardinier, NY, 1877. An unusual and different type of fortune telling deck that has a patent and copyright dated 1877. The description of the deck is quoted from the box: "Every picture is a burlesque on the true meaning of each card, but nothing to offend or shock the most fastidious, being suitable for any parlor or private circle, and can be used for playing all games the same as with ordinary cards. These are the only Comic Playing Cards where Kings, Queens and Jacks are the same as in all ordinary cards, hence do not have to be learned anew, as is always the case when they have been distorted into fantastic figures". The Joker states, "This card may be used as a Joker, but should be withdrawn from the pack in Fortune Telling."

FT9

FT9 YE WITCHES FORTUNE TELLING CARDS

USPC, 1896. This deck, besides being a novel fortune telling deck, is also considered a partial transformation deck (T12). Each card has a different sketch in cartoon style and there is no writing on the faces. There is an accompanying booklet that describes the layout and readings. The court cards, although redesigned, are quite standard.

FT10

FT10 EGYPTIAN FORTUNE TELLING AND TRICK CARDS

H.A. Rost Printing Co., NY, 1898. E. Rost designed this deck with a different fortune printed on all four sides of each card. An accompanying booklet included "Secret Rules for the Novice". The instructions for the tricks are such that they could be performed with any regular deck.

FT11

FT11 NILE FORTUNE TELLING

USPC, 1897. The first of a standard card series that were designed for fortune telling by adding fortunes printed in the margins. This Nile deck had two different fortunes, one on each of the top and bottom margins.

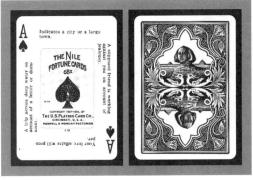

FT12

FT12 NILE FORTUNE TELLING II
USPC, 1904. This is the almost the same deck as FT11, with the same instruction booklet. It was issued under a new copyright and had a new set of fortunes, but this time on all four margins of every card. The creation of a new edition seems to indicate the success of this type of playing cards.

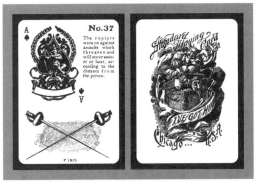

FT13

FT13 GYPSY WITCH CELEBRATED FORTUNE TELLING CARDS
Madame Le Normand, copyright 1903 by Frederick J. Drake Co. First published by Home Game Co., 1903, then by Standard Playing Card Co., 1904. This deck has been published continuously ever since using the same faces. More recent editions have a witch on the backs replacing the original river scene.

FT14

FT14 ASTROLOGICAL FATE
USPC, 1908. This deck represents an attempt to revert to the tarot for fortune telling, but in a modernized way. The deck consists of 67 cards, the normal 52, twelve Zodiac Cards and three special Planetary Cards, The Sun, The Moon and Mars. The Astrological Fate deck came with a book of instructions, a booklet explaining its inspiration from the early tarot and a large chart on which to spread the cards. Pictured are The Ace of Spades, and a Zodiac card (Gemini). All of the pip cards have fortunes in the margins and a symbol in the right hand corner. The courts are standard.

FT15

FT15 REVELATION FORTUNE TELLING
USPC, 1919. A deck quite similar in design and purpose to the Nile (FT11 and FT12). It serves as evidence of the continued popularity of this type of fortune telling deck.

FT16

FT16 MILITARY FORTUNE TELLERS
H.V. Loring, Chicago, 1918. This deck has also been listed as W27 and NS7 but this is its main usage.

FT17

FT17 WHEEL OF FORTUNE
Elizabeth B. Leonard, Providence, RI, c1920. This is an interesting and unusual type of fortune telling deck that has two messages on each card. Both messages are written on the rim of the wheel, facing in opposite directions.

FT18

FT18 LET'S TELL FORTUNES
Fortune Playing Card Co., Spring Valley, NY, 1941. The only thing appearing on the face of each card, besides the indices, are two messages, each facing in a different direction. This deck is unusual as it consists of only 48 cards with the deuce of each suit being omitted.

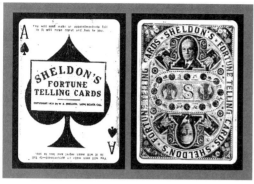

FT19

FT19 SHELDON'S FORTUNE TELLING
c1930. The fortunes are on both ends of this deck.

FT22

FT22 WIZARD CARDS
NYCC, 1900. This is a curious deck made for the Wizard Card Co., 1368 Broadway, New York. It shows two printed fortunes on each card. The cards are standard and have the NY50 Ace of Spades.

FT20 CARDS OF FATE (not pictured)
J.H. Singer, NY, c1885. An early Le Normand style deck which came with 26 cards and the instructions. It was packaged in a very colorful game box. It is not pictured as the cards are the same as FT4.

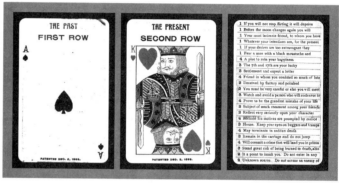

FT23

FT23 MME. DOERFLINGER PATENT FORTUNE TELLING CARDS
Eureka Chemical and Mfg. Co., Lacrosse, Michigan, 1893. This deck has only 24 cards consisting of nine to Ace in each suit, plus an index card.

FT21

FT21 CABALISTIC PLAYING CARDS
Albert Mudge & Son, Printers, 34 School Street, Boston, MA, 1872. A wonderful deck with single ended colored courts and fortunes printed in all the margins. There are 52 cards and four fate cards, no joker being issued with this deck. There is an accompanying Cabalistic booklet.

FT24

FT24 LADIE LUCILE'S FORTUNE CARDS
New York, c1900. A fortune telling deck of 36 cards patterned on cards previously published in Germany and England.

EXPOSITION AND WORLD'S FAIR

The decks in this section were mainly issued for sale at a particular fair or exposition as souvenirs. It is clear that certain of the ones produced by USPC were actually issued after the fairs referred to on the cards showing the medals won by the company. They have been listed here for consistency with the original Encyclopedia. Exposition decks are much sought after and highly collectible. Although not as scarce as other types, these decks will often bring higher prices because playing card collectors are competing with World's Fair collectors to obtain the available specimens.

SX3   SX4

SX3 CENTENNIAL PLAYING CARDS
The Continental Card Co., Philadelphia, 1876. This standard deck (also listed as U18a) was issued to be sold at the exposition with only the backs, wrapper and Ace of Spades mentioning the event.

SX4 ROYAL PLAYING CARDS
NYCC, c1890. It was originally thought that this beautiful deck was issued for the Paris Exposition in 1878, and it was therefore listed in this category. It has subsequently been demonstrated (refer Clear The Decks, Vol. VII, No.2) that it was published in the mid-1890s as a deck commemorating the Triple Alliance of Germany, Austria and Russia (subsequently replaced by Italy) with suits dedicated to these four powers. It has therefore been listed in the political chapter as P22.

SX1   SX1a

SX1 CENTENNIAL PLAYING CARDS
Victor Mauger, NY, 1876. These are the first known American cards that commemorate a fair or exhibit. They have standard faces but are unique in two different ways. They are called 'Quadruplicates' because they feature four corner indices. This was unusual as four way indices, which became popular on European cards, were rarely used on U.S. cards. This was also very innovative as at that time indices were only beginning to be used on playing cards. The deck also features four different colored suits - the spades are black; the hearts, red; the diamonds, yellow; and the clubs, blue. The Joker, used also in their standard cards (Chapter 3), shows George Fox, a popular entertainer of the day.

SX1a CENTENNIAL PLAYING CARDS
Victor Mauger, NY, 1876. This deck (also listed as U19b) appears to be another centennial deck as the Ace of Spades features the motto and the date of the centennial encircling the spade pip. The court cards are two-way and the suits are in the traditional colors rather than the four-color version of SX1.

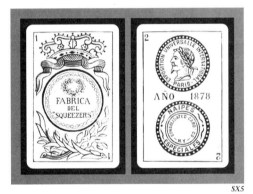

SX5

SX5 FABRICA DEL SQUEEZERS
NYCC, 1878. This deck represents one of the first attempts by NYCC to capture a portion of the foreign market. Note the tiny indices in the corners of each card. The deck won a prize at the Paris Exposition and perhaps was sold there.

SX2 CENTENNIAL PLAYING CARDS
Andrew Dougherty, New York, 1876. This deck was issued with backs featuring an oval portrait of George Washington surrounded by a patriotic motif and the words "America's Centennial, July 4th, 1876". This was a triplicate deck with a special back for the centennial.

SX2

SX6

SX6 COLUMBIAN EXPOSITION SOUVENIR PLAYING CARDS
G.W. Clark, Chicago, 1893. The backs feature 'The Landing of Columbus'. The faces have round colored sketches of the fair and miniature playing cards in the upper left and lower right corners. The value of the card is shown at the top and bottom of every card. The Joker shows two seals and the words "Columbian Souvenir Playing Cards".

SX7

SX7 COLUMBIAN EXPOSITION SOUVENIR PLAYING CARDS
Winters Art Litho. Co., Chicago, 1893. The backs feature the three ships of Columbus in pink or blue. Faces depict scenes and buildings of the fair. Bust portraits on the upper left corner of each court card show Ferdinand on the Kings, Isabella on the Queens and Columbus on the Jacks.

SX8   *SX8a*

SX8 COLUMBIAN EXPOSITION SOUVENIR PLAYING CARDS
Winters Art Litho. Co., Chicago, 1893. The backs and scenes of these cards are identical to those of SX7. The bust portraits however are in the upper right hand corner with Colonel George R. Davis, the President of the fair on the Kings, Mrs. Potter Palmer on the Queens with Columbus remaining on the Jacks.

SX8a COLUMBIAN EXPOSITION SOUVENIR PLAYING CARDS
World's Fair Souvenir Card Co., 1892. This deck was issued a year earlier than SX7 and SX8. The back design is slightly different but with the same motif. The bust portraits of Colonel Davis, Mrs. Potter and Columbus are on the left side of the card.

SX9 COLUMBIAN EXPOSITION SOUVENIR PLAYING CARDS
(not pictured)
Winters Art Litho. Co., Chicago, 1893. This deck was listed in the original Encyclopedia as having the same scenes as the previous listings and courts without the bust portraits. In addition the backs were described as having a large Winter's medallion as the central theme. All examples of this back have been discovered to be Kings of Hearts and it is now known that the cards were included in decks of SX8 as a promotional item.

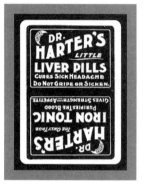

SX10

SX10 COLUMBIAN EXPOSITION
Winters Art Litho. Co., c1893. The faces of these cards are identical to SX8 but they are advertising cards for the Dr. Harter Medicine Co., Dayton, Ohio. The backs advertise 'Dr. Harter's Little Liver Pills' and 'Iron Tonic'. The same ad appears on the Joker that states "The Only True Iron Tonic".

SX11

SX11 COLUMBIAN EXPOSITION
Winters Art Litho. Co., c1893. This deck is also identical to SX8 except that two cards, the Queen of Clubs and the Jack of Diamonds differ. They show, respectively, an advertisement for the Hayner Company and a price list.

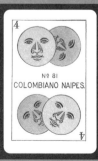

SX12

SX12 COLUMBIANO NAIPES #81
USPC, 1893. This is a beautiful Spanish deck that was issued for the exposition. There are South American Indians pictured on all of the courts. In place of the normal Spanish suit signs Coins are replaced by Suns, Cups by Earthen Bowls, Swords by Arrows, and Clubs by Indian War Clubs. This was one of several decks for the Spanish market issued by USPC. Although this deck was made and sold until 1919 it is very scarce.

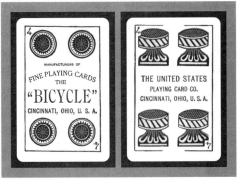

SX13

SX13 NAIPES ESPECIALES #95
USPC, 1893. Another USPC Spanish suited deck, perhaps for sale at the fair, traded on the Bicycle brand and used regular Spanish suit signs.

SX14

SX14 LOS LEONES NAIPES #71
USPC, 1893. This deck with Spanish suit signs was USPC's attempt to attract the Mexican market. It was issued as a 40 card deck for the game of 'Hombre' instead of as a regular 48 card deck. This deck was first issued at the Columbian Exposition, but was printed for many years into the 20th century.

SX13a

SX13a NAIPES ESPECIALES #95
USPC, c1904. A different version not using the Bicycle name was manufactured later and the cards emphasised the medals won at the various world's fairs.

SX15

SX15 MIDWINTER INTERNATIONAL EXPO
Winters Art Litho. Co., Chicago, 1894. This was made as a continuation of the successful cards at the 1893 fair. The backs feature State seals of California, Oregon and Washington along with the Seal of the Exposition. The faces have color sketches of the fair and bust portraits of Uncle Sam, Liberty, etc. in the upper right-hand corners. The Joker shows a walking brown bear.

SX13b

SX13b NAIPES BICICLETA #810
USPC, 1904. Another version of SX13a stressing the medals won in Chicago, St. Louis, etc.

SX16

SX16 MIDWINTER EXPOSITION
Winters Art Litho. Co., 1894. These cards were issued as an advertising deck for Enterprise Brewing Co., San Francisco. They are identical to SX15 except for two cards. The Jack of Diamonds and Queen of Spades have brewery ads and say "compliments of Enterprise Brewing Co.", which leads one to believe they were given away or sold at the Enterprise exhibit at the fair.

SX17   SX18

SX17 PARIS EXPOSITION SOUVENIR PLAYING CARDS

Tom Jones, Denver, 1900. The backs feature a statue, "The Modern Parisians", which stood at the grand entrance to the exposition. The faces have oval photo scenes. The deck is listed in this chapter, even though it was a foreign fair, as it was made by USPC and likely published by Tom Jones for sale mainly in the United States.

SX18 PAN AMERICAN EXPOSITION SOUVENIR PLAYING CARDS

Pan American Souvenir Playing Card Co., Buffalo, 1901. The backs feature a relief map of North and South America joining hands. The faces have a different oval photo scene of the exposition on each card.

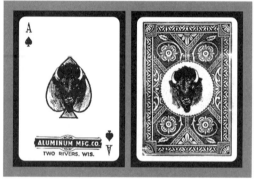

SX19

SX19 ALUMINUM PLAYING CARDS

Aluminum Mfg. Co., Two Rivers, Wis., 1901. These are the first aluminum cards known. They have standard faces with the American Bison, the symbol of the Pan American Exposition, on the back, the Ace of Spades and the Joker. They were sold at the company's exhibit.

SX20 ALUMINUM PLAYING CARDS

Aluminum Mfg. Co., Two Rivers, Wisconsin, 1904. This deck was made for the St. Louis World's Fair. The backs feature a circular portrait of Thomas Jefferson and the words "World's Fair, St. Louis, USA 1904". The face of each card shows a different sketch of a scene at the fair. It was issued in a special aluminum case inscribed as a souvenir of the fair.

SX20

SX21

SX21 ST. LOUIS WORLD'S FAIR (LOUISIANA PURCHASE) SOUVENIR PLAYING CARDS

Samuel Cupples Envelope Co., St. Louis, Mo., 1904. The backs feature the official seal of the exposition, with profiles of Napoleon and Jefferson centered in an eagle that holds a hemisphere in each claw. Oval photo scenes are on the face of each card.

SX22

SX22 PHILIPPINE SOUVENIR PLAYING CARDS

The Philippine Photograph Co., 1904. These were made expressly for the St. Louis World's Fair, and were sold exclusively at the Philippine exhibit. The only mention of the fair appears on the box. The back features a man on a long-horned buffalo doing farm work. The oval photo scenes on the face of each card depict life in the Philippines.

SX23

SX23 JAMESTOWN EXPOSITION SOUVENIR PLAYING CARDS

Old Dominion Paper Co., Norfolk, 1906. The backs portray "The Baptism of Pocahontas". Each face has a different photo scene of the exposition, or a sketch of one of the founders of Jamestown colony. The Ace of Spades shows the coat of arms of Captain John Smith. This deck was subsequently issued as Virginia Souvenir Playing Cards and is also listed as S81.

SX24

SX24 ALASKA YUKON PACIFIC EXPOSITION SOUVENIR PLAYING CARDS

USPC Co., 1909. The backs feature a scene of Mt. Rainier plus the Seal of the exposition in each corner. The faces have different oval photo scenes of the fair and of the City of Seattle, which was the exposition site.

SX25

SX25 PANAMA PACIFIC EXPOSITION SOUVENIR PLAYING CARDS

San Francisco, 1915. This exposition celebrated the opening of the Panama Canal. This deck is, however, not the usual souvenir type with scenes on each card. It has standard faces with the backs showing the 'Tower of Jewels', a feature of the fair.

SX26

SX26 CENTURY OF PROGRESS

Western P.C. Co., 1933. Chicago World's Fair Souvenir Cards featuring a globe rotating on its orbit with the date 1933. Each face has a different photo scene of the Chicago World's Fair. The Joker shows a gorilla of a million years ago.

SX27

SX27 CENTURY OF PROGRESS

Western P.C. Co., 1934. This is a later edition of SX26 with the backs dated 1934. The Joker is identical, but the scenes on the faces are different.

SX31

SX31 NEW YORK WORLD'S FAIR SOUVENIR PLAYING CARDS

1939. The backs feature the 'Trylon & Perisphere' symbols of the fair. The faces have oval scenes of the many attractions.

SX28

SX28 CENTURY OF PROGRESS

Arrco PCC. There were many decks, some with scenic faces and others with standard ones, made by Arrco for this fair. The backs were used interchangeably with standard and scenic faces and we have therefore combined all Arrco decks under this number and eliminated SX29 and SX30.

SX32

SX32 NEW YORK WORLD'S FAIR SOUVENIR PLAYING CARDS

1939. The backs also feature the Trylon and Perisphere. This deck however, has standard faces. All other standard faces, of which there are many, should be given this reference.

SX33

SX33 NEW YORK WORLD'S FAIR

1939. The backs feature the General Motors exhibit, 'Futurama' and the deck has the same scenic faces as SX31.

SX34 TEXAS CENTENNIAL

1936. Issued to celebrate Texas' entry into the union. The back has a flag flying over the Alamo.

SX34a   SX34b

SX34a TEXAS CENTENNIAL

1936. A similar deck, but this back features the Texas map.

SX34b TEXAS CENTENNIAL

1936. Another version features the Texas star on the back.

SX35

SX35 NEW YORK WORLD'S FAIR

Stancraft, 1964-65. The backs feature the Unisphere, the symbol of the fair.

SX36

SX36 NEW YORK WORLD'S FAIR

KEM PCC, 1939. A special issue made by this quality maker of plastic cards especially for the fair.

SX37

SX37 THE SESQUICENTENNIAL

King Press, 1926. This deck was published by Young & Rudolph of Philadelphia in celebration of the 150th anniversary of the USA.

SX38

SX38 THE COLUMBIAN EXPOSITION

Perfection Card Co., New York, 1893. This standard deck was issued for the Columbian Exposition in 1893. It has also been listed in Chapter 9 as PU12. The top of the box has '1492', the bottom '1893' and the Joker is a portrait of Columbus.

WIDE SOUVENIRS OF STATES, CITIES AND NATIONAL PAR

The playing cards in this section are the wide souvenir issues that date from about 1890 to the early 1930s. They feature photographic scenes, on the face of each card, of the area that they represent. Some of these decks are rare, but as most of them had large print runs and fall within a category almost exclusively of interest to playing card collectors, they can usually be purchased quite reasonably. There are certain topics however, for example Indian people and artefacts, which have interest to collectors in other fields thereby raising the demand and consequently the value of these decks.

Narrow souvenir decks are not listed in this Encyclopedia. There are literally thousands of these printed from the late 1920s to this day and they are of relatively limited value and interest to deck collectors.

In some cases the actual manufacturer of a particular wide souvenir deck, despite lengthy research, could not be determined. For example, it is known that The United States Playing Card Co. was the maker of many decks attributed to Tom Jones, but we cannot be sure that at one time a different manufacturer was not used. The Standard Playing Card Co. of Chicago was also the manufacturer of several decks credited to others. Many decks are therefore listed under the name of the distributor rather than the manufacturer as no clue to the maker could be found on the specimens available for examination.

Wide souvenir decks have survived for They were souvenirs - often tucked awa_____ to be reviewed as a remembrance of a trip or vacation. To many, they were of sentimental value and therefore rarely used. They also lasted well, generally being the top quality grade of the manufacturer, and with a very few exceptions offered in sturdy, heavy cardboard slipcases. Finally, they were very difficult to play with, as they did not have the familiar court and pip cards. It is not surprising therefore, when purchasing a deck of early souvenir cards, that one will slip open a case and be delighted to find a mint specimen with shiny gold edges.

The listing of decks in this chapter is probably complete, however an omission is by no means impossible. In fact, there are several additions since the publication of Volume I of the original Encyclopedia. All wide souvenir decks were issued with jokers, in many cases the jokers representing the fifty-third view promised on the case. The decks also often came with from one to three additional cards describing the views, mapping the area or providing general information about the topical subject.

With just a few exceptions, United States souvenir decks have one of five different types of photo scenes on their faces. To simplify identification, we picture one example of each type below.

TYPE A   TYPE B

TYPE C   TYPE D

TYPE A
The earliest type starting about 1890 was printed in either black and white or sepia oval-shaped scenes with no outline.

TYPE B
This is also a very early type with the addition of a fine ornate outline for the scene. The oval photos were still in black and white or sepia.

TYPE C
Starting in 1901, the monotone oval scene was introduced. These photos have an outline of the same color. It is not unusual to find that each suit is printed in a different color or tone. A large number of the decks in this section fall into this category.

TYPE D
Several of the decks in this section are of this type manufactured by Standard Playing Card Co. and bearing the name of either Bosselman or Standard. They have greyish photo scenes with distinct edges. The titles are printed beneath the oval scenes.

TYPE E

The most recent kind of photo scenes are the round cornered, rectangular shaped variety which first appeared around 1915.

TYPE E

The following four decks, showing Indians and Indian life are of significant interest to many other collectors and are, consequently, difficult to find and more expensive then many other souvenir decks. S4, picturing a Hopi Indian boy, is especially rare.

There are some differences found in the photos used within certain decks. We are not going to include these variations as many souvenir decks were made over a long period of time and it would not be unusual to find that the publishers had substituted different photographs on certain cards. Generally, we have continued Gene Hochman's style of showing only the back for each listing.

S3   *S3A*

S3 AMERICAN INDIAN SOUVENIR CARDS
Lazarus and Melzer, 1900. The back is a colorful painting of an Indian rug in black, gold and red. The faces show oval photo scenes (Type A) featuring 'A series of Artistic Indian Views'. The King of Hearts is Kopeli, Chief of the Snake Priests. The Joker has an Indian standing in a meadow.

S3a AMERICAN INDIAN SOUVENIR CARDS
P. Lazarus, 1903. This deck with the same back as S3, has a completely different set of Type A photos. The pictured Joker shows a woven Indian basket.

S1 ALASKA SOUVENIR PLAYING CARDS
Edward H. Mitchell, 1900. The back features the Seal of the District of Alaska flanked by two totem poles. The faces show oval photo scenes (Type A) of the Territory of Alaska.

S1

S4   *S5*

S4 AMERICAN INDIAN SOUVENIR PLAYING CARD
Fred Harvey, Kansas City, 1908. The back features a picture of a young Indian sitting on the lowest rung of a ladder. The title 'A Hopi Boy' is at the lower left. Faces show oval photo scenes (Type B) of Indians and Indian Life.

S5 INDIANS OF THE SOUTHWEST SOUVENIR PLAYING CARDS
Fred Harvey, c1912. This back has a circular design in the center of the card highlighting an Indian god with feathered arms. The faces show oval photo scenes (Type C) of Indians and Indian life.

S2 ALASKA SOUVENIR PLAYING CARDS
Puget Sound News Co., c1910. The back features a scene of Alaska with foliage and two totem poles. The faces show rectangular photo scenes (Type E). This deck was sometimes sold in a brass holder bearing the Seal of the District of Alaska.

S2

S6

S6 HISTORIC BOSTON & VICINITY SOUVENIR PLAYING CARDS

Chisolm Bros., Portland, Me., c1900. The back has a large oval with the Bunker Hill Monument in the center. The title is in black on a white band below the picture. Oval photo scenes (Type A) show views of the area.

S7 HISTORIC BOSTON & VICINITY SOUVENIR PLAYING CARDS

(not pictured)
Chisolm Bros., c1910. This back is the same as S6 except that the oval is outlined in gold and the title is in gold print. The cards have the same oval photo scenes (Type C) as the earlier issue.

S8

S8 BUFFALO & NIAGARA FALLS SOUVENIR PLAYING CARDS

S.O. Barnum & Son, Buffalo, c1905. The back comes in either green or red and depicts a river scene. The faces show oval photo scenes (Type C) of the area.

S9

S9 BUFFALO & NIAGARA FALLS SOUVENIR PLAYING CARDS

Niagara Playing Card Co. (also issued by S.O. Barnum), c1910. The back, entitled 'Maid of the Mist', portrays an Indian maiden in a canoe paddling over the waters of Niagara Falls. The faces show oval photo scenes (Type C) of the area.

S10

S10 THE BURRO SOUVENIR PLAYING CARDS

H.H. Tammen Curio Co., Denver, 1904. The back has a sketch of a burro's head titled 'They Call Me Satan'. The faces show oval photo scenes (Type A) all portraying burros at work and at play. This is another very hard to find deck.

S11   S12

S11 CALIFORNIA SOUVENIR PLAYING CARDS

R.J. Waters, San Francisco, 1898. The monotone red or blue back shows the California State Seal surrounded by four white poppies. The faces have oval photo scenes (Type A) on each card.

S12 CALIFORNIA SOUVENIR PLAYING CARDS

R.J. Waters, 1898. This back has the California State Seal on a grey background surrounded by orange poppies. Above the Seal are the words 'State Flower' which measure one-half inch and below reads, 'California Poppy'. The faces show oval photo scenes (Type A) that are different from those in S11.

S13   S14

S13 CALIFORNIA SOUVENIR PLAYING CARDS

M. Rieder, 1907. Rieder was probably the successor to Waters as their cards are very similar. The back is almost identical to S12 except the poppies do not have brown centers and the words 'State Flower' measure five-eighths of an inch. The faces show different oval photo scenes (Type C).

S14 CALIFORNIA SOUVENIR PLAYING CARDS

M. Rieder, 1913. This back is covered with large orange poppies with green stems. The cards have a light green border with a dark green background. The flowers extend past the border on all sides to the ends of the cards. The faces have Type C oval photo scenes.

S15

S15 CALIFORNIA SOUVENIR PLAYING CARDS

M. Rieder, 1907. The back shows four oranges surrounded by orange blossoms. The faces have different oval photo scenes (Type C).

S16   *S17*

S16 CALIFORNIA SOUVENIR PLAYING CARDS
M. Rieder, 1912. The back is the same as S15 but with a light green border around the cards. The faces are almost the same although some photos are different.

S17 CALIFORNIA SOUVENIER PLAYING CARDS
M. Rieder, 1907. This back has the Mission Bell in a brown frame flanked by poinsettias. There are Type C oval photo scenes which are the same as either S13 or S16.

S18   *S19*

S18 CALIFORNIA SOUVENIR PLAYING CARDS
M. Rieder, 1913. The back again features the Mission Bell but there is a light green border and the bell is in a smaller gold frame. Within the frame is the title 'Old Mission Bell'. The faces show completely new oval photo scenes (Type C).

S19 SOUVENIR PLAYING CARDS OF CALIFORNIA
Bullock's, c1920. This edition was made for Bullock's Department Stores in Los Angeles, the Emporium in San Francisco and other smaller stores. The back shows an orange grove with the Pacific in the foreground and snow capped mountains in the background. The faces show rectangular photo scenes (Type E).

S20 COLUMBIA RIVER HIGHWAY SOUVENIR PLAYING CARDS
The Souvenir Playing Card Co., Portland, c1920. The back depicts a scene of high narrow falls going down a mountain side. The title 'Multnomah Falls' is in gold print. The faces show rectangular photo scenes (Type E).

S20

S21   *S22*

S21 FLORIDA SOUVENIR PLAYING CARDS
Tom Jones (USPC), c1900. The back has a single shaded orange blossom in green tones on either a yellow or orange background. The faces show oval photo scenes (Type B).

S22 FLORIDA EAST COAST SOUVENIR PLAYING CARDS
Interstate News Co., New York, c1900. Some believe this deck should be listed with the Railroad Decks, but there is no indication on the cards or box as to whether it is Florida East Coast R.R. or simply the East Coast of Florida. The back shows the Seal of Florida surrounded by Orange Blossoms. The faces have oval photo scenes (Type C).

S23 GRAND RAPIDS, MICHIGAN SOUVENIR PLAYING CARDS
Dickinson Bros., Grand Rapids, 1910. The back is divided diagonally by a pole having a 'Furniture City' banner at each end. Multiple white outlined triangles with GRM monograms fill the background. This deck differs from the other decks in this section as it is made on cheaper, heavier stock and has a different type photo scene. It was likely produced in relatively small quantities, as it is very difficult to find.

S23

S24   *S25*

S24 GREAT LAKES SOUVENIR PLAYING CARDS WESTERN DIVISION
USPC Co., 1909. The back features an oval portrait of St. Isaac Jacques surrounded by a scene of two Indians paddling a canoe with tall trees in the background. The faces have oval photo scenes (Type C) of the area.

S25 GREAT LAKES SOUVENIR PLAYING CARDS EASTERN DIVISION
USPC Co., c1910. This back features an Indian design with an Indian woman with long black braids in typical Indian dress sitting on a blanket. It is believed she is Nokomis, the mother of Hiawatha. The faces have Type C oval photo scenes.

S26

S26 GREAT SOUTHWEST SOUVENIR PLAYING CARDS

Fred Harvey, 1901. The back has a scene of Indians and horses on the plains. In the background a train is crossing the desert against a clear blue sky and is emitting a short smoke stream. Oval photo scenes (Type C) of Indians and the area are on the face of each card. The Ace of Spades is entitled 'Harvest Dancers'.

S26a

S26a GREAT SOUTHWEST SOUVENIR PLAYING CARDS

Fred Harvey, 1901. A slightly different version of this deck has a somewhat longer smoke stream and different Type C photos. The Ace of Spades is 'Lookout'.

S27

S27 GREAT SOUTHWEST SOUVENIR PLAYING CARDS

Fred Harvey, 1910. This is a later issue of the same deck. The back is almost identical but in this version the sky is cloudy and the train emits an even longer smoke stream. The Type C photo scenes are different with the Ace of Spades being 'Jicarilla Chief'.

S28   S29

S28 HAWAIIAN SOUVENIR PLAYING CARDS

Wall, Nichols Co., Honolulu, 1901. The back has a statue of King Kamehameha I surrounded by a gold frame. The faces show grey oval photo scenes (Type A) of early Hawaii.

S29 HAWAIIAN SOUVENIR PLAYING CARDS

Wall, Nichols Co., 1901. This deck although dated 1901, is probably a later edition. The back has the same statue as S28 but a fancy green frame has replaced the gold one. The faces have oval photo scenes with yellow backgrounds (Type C) and are the same scenes as those of S28.

S30

S30 INTER-MOUNTAIN SOUVENIR PLAYING CARDS

USPC Co., c1900. The back has a slanted oval at the center of the card with a scene of a prospector walking his mule along a mountain road. The faces depict oval photo scenes (Type C) in the Rockies.

S31

S31 INTER-MOUNTAIN SOUVENIR PLAYING CARDS

USPC Co., c1900. This back has a scene of a covered wagon train moving along the side of a mountain. It is in a gold, outlined circle in the center of the card. The faces have the same oval photo scenes as S30.

S32

S32 MAINE SOUVENIR PLAYING CARDS

Chisolm Bros., c1900. The back shows a black and white photo of the State House in Augusta, Maine set in a gold bordered, rectangular frame with the title printed in gold below the photo. In the two upper corners are Seals of the State of Maine. The faces have Type C oval photo scenes.

S33

S33 MONTANA AND YELLOWSTONE SOUVENIR PLAYING CARDS

The Photo Card Co., Butte, c1898. The back features mountains and a banner reading 'Oro y Plata' within a circle with mining scenes and prospectors above and below. The faces have Type A photo scenes.

S34 MOUNTAIN STATES SOUVENIR PLAYING CARDS
Gray News Co. (USPC), 1914. The back shows the sun setting behind dark clouds with its reflection on the rippling water. The faces have oval photo scenes (Type C).

S34

S39

S40

S35 NEW ORLEANS AND GULF SOUVENIR PLAYING CARDS
USPC Co. c1900. The back has pale pink magnolia blossoms forming a frame outlined in gold of a plantation scene. The faces have Type C oval photo scenes.

S35

S39 NEW YORK CITY VIEWS
A. C. Bosselman, New York, Standard PCC, c1910. The back, in either green or red, has a monotone river bank scene used for several Bosselman decks. The faces show oval photo scenes (Type D) of New York City.

S39a NEW YORK CITY VIEWS *(not pictured)*
Standard PCC, Chicago, c1910. Although Standard made the Bosselman deck they also introduced their own version. The back was the same and the photo scenes were the same Type D ones but with many different views.

S40 NIAGARA FALLS SOUVENIR PLAYING CARDS
Niagara Playing Card Co., Buffalo, 1901. The back is in blue bas-relief and shows an Indian on horseback attacking a buffalo. The deck has a dark blue border. The faces show oval photo scenes (Type A) of Niagara Falls.

S41 NIAGARA FALLS SOUVENIR PLAYING CARDS *(not pictured)*
Niagara Playing Card Co., Buffalo, 1901. This back is the same as S40 except that it has the appearance of a sketch rather than of a photo. The border is a bamboo type arrangement. The color is either greyish blue or mauve. The faces have the identical Type A oval photo scenes. Both decks date 1901, but it is not known which was issued first.

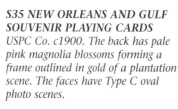

S36　　　　*S37*

S36 NEW YORK CITY SOUVENIR PLAYING CARDS
USPC Co., c1900. The faces of this deck show oval photo scenes (Type C) of New York City and the back features the Statue of Liberty without a border

S37 NEW YORK CITY SOUVENIR PLAYING CARDS
USPC Co., c1915. This back has the Statue of Liberty surrounded by a white border. The faces again show oval photo scenes of New York City but of a later vintage than those of S36.

S42 NORTH SHORE MASSACHUSETTS SOUVENIR PLAYING CARDS
D.R.M. Speciality Co., Gloucester, 1906. The back has a large fishing schooner with two masts, under full sail in a gold highlighted rectangle within a border of codfish and seaweed. The faces have oval photo scenes (Type C).

S42

S38 NEW YORK CITY AND HUDSON RIVER SOUVENIR PLAYING CARDS
Edward Mitchell, San Francisco, c1911. The back depicts Father Knickerbocker in his tri-cornered hat in an oval frame. There are oval photo scenes (Type A).

S38

S43 O'CALLAGHAN'S SOUVENIR PLAYING CARDS
(City of Chicago), John J. O'Callaghan, Chicago, 1930. The back on this deck has a large rectangular etching of the Civic Opera House beyond the Chicago River. The border is either red or green. There are rectangular photo scenes on the face of each card with triplicate type insets in two opposite corners showing the value of the card.

S43

S44

S44 OREGON SOUVENIR PLAYING CARDS

Edward H. Mitchell, San Francisco, 1901. The back has an oval frame containing Mt. Hood reflected in water on a deep green and gold background. The faces have Type A scenes.

S45

S45 PANAMA SOUVENIR PLAYING CARDS

Inaugural Edition, USPC 1915. This deck was issued at completion of construction of the Canal. It was sold at the Panama Pacific Exposition as well as to visitors traveling through the Canal. The back depicts a woman in a long white gown, with long hair, standing on the shore facing the sunset. The faces show Type C photo scenes along the canal.

S46   S47

S46 PANAMA SOUVENIR PLAYING CARDS

USPC, 1908. This deck, issued at the commencement of work on the Canal, was printed in 1908. The back has a scene of a road lined with royal palm trees. The faces have oval photo scenes (Type C) with multi-lined captions. There are several extra cards including 'Facts Regarding the Panama Canal 1908' and a two-sided map card showing Panama in 1908 and on the reverse, 'Panama at the completion of the Canal'.

S47 PANAMA SOUVENIR PLAYING CARDS

USPC, 1910. This back has a scene enclosed in a gold oval frame entitled, in gold, 'Chagres River View'. The faces have oval photo scenes (Type C). The four suits show different aspects of Panama and the work on the Canal. Diamonds (green tint) are devoted to the Canal Zone and life under United States rule. Clubs (buff tint) give us an idea of the excavation and work along the Canal. Spades (pink tint) show the construction of the massive locks and dams. Hearts (blue tint) portray life in Panama.

S48   S49

S48 PANAMA SOUVENIR PLAYING CARDS

USPC, 1923. The back features a large black ship steaming through the canal that reflects the red/gold sky in the water. The faces have oval photo scenes (Type C) and the heart and diamond suits are identical to S47 but the black suits showing the construction have been updated.

S49 PANAMA SOUVENIR PLAYING CARDS

USPC, 1926. The back has a scene of a black and white ship moving through the finished canal on blue water. The faces show oval photo scenes (Type C) of the canal.

S50

S50 PANAMA SOUVENIR PLAYING CARDS

USPC, 1926. The back on this version has a black and white ship passing through the canal. An old galleon is pictured in the smoke rising from the stacks. The borders are white, pink, peach, yellow, or green. The faces have rectangular Type E photo scenes.

S50a PANAMA SOUVENIR DECK *(not pictured)*

An earlier issue of this deck with oval (Type C) views was also made by USPC.

S51

S51 PITTSBURG SOUVENIR PLAYING CARDS

W.J. Gilmore, Pittsburg, 1901. The back has the flag of Pittsburg surrounded by black and gold symbols. The faces show oval photo scenes (Type A) of the city.

S52

S52 VIEWS OF PITTSBURG

May Drug Co. (USPC), c1905. This back shows the Blockhouse at Fort Duquesne in a small circle flanked on either side by the Seal of the City of Pittsburg. Above and below are pennants of the University of Pittsburg and of Carnegie Tech. The faces have Type A oval photo scenes of the city.

S53

S53 PORTLAND BY THE SEA SOUVENIR PLAYING CARDS

Chisolm Bros., Portland, Maine, c1910. The back has a photo of the Portland Lighthouse in a small diamond shaped frame. Above and below the photo are Seals of the State of Maine and the City of Portland on a background of pinecones. The faces show oval photo scenes (Type C).

S54 PORTLAND AND COLUMBIA RIVER SOUVENIR PLAYING CARDS

USPC, 1901. The back has an oval picture of Mt. Hood surrounded by a background of pink roses. The faces have Type C scenes.

S54

S55 RHODE ISLAND SOUVENIR PLAYING CARDS

USPC, 1910. The back has a small rectangular photo of the State Capitol with the Seals of the State of Rhode Island surrounded by a background of sprays of goldenrod. The faces show oval photo scenes (Type C) of items of interest in the state.

S55

The following four decks are all listed as published by Tom Jones in 1899 as souvenirs of the Rocky Mountains.

S56   *S57*

S56 ROCKY MOUNTAIN SOUVENIR PLAYING CARDS

Tom Jones, 1899. The first back has a single purple columbine on either a beige or yellow background. The faces have oval photo scenes (Type B).

S57 ROCKY MOUNTAIN SOUVENIR PLAYING CARDS

Tom Jones, 1899. The second back is reversible with a single columbine on a pink background. A scalloped border of green leaves and columbines surrounds the design. The faces show oval photo scenes (Type B).

S58   *S59*

S58 ROCKY MOUNTAIN SOUVENIR PLAYING CARDS

Tom Jones, 1899. This back has a reversible design of purple columbines and green leaves. In fine print are the words 'Colorado State Flower - Columbine'. The faces have oval photo scenes (Type A).

S59 ROCKY MOUNTAIN SOUVENIR PLAYING CARDS

Tom Jones, 1899. This back has a bouquet of purple and white columbines and green leaves. The faces have the same oval photo scenes to those of S58.

S60

S60 ST. JOSEPH, MISSOURI SOUVENIR PLAYING CARDS

USPC, 1907. The back has a small orange map of the United States with St. Joseph in the center. It is surrounded by a circle upon which is written in gold 'St. Joseph Missouri Western Gateway'. The dates 1851 and 1907 are in the corners. The faces have Type C oval photo scenes.

S61

S61 ST. LOUIS SOUVENIR PLAYING CARDS
Meyer Bros. Drug Co., St. Louis, 1901. The back features a painting of Louis XVI in a white wig, ermine robes, and other royal regalia. The cards have either a green or black border. The cards come with oval photo scenes (Type A).

S62

S62 TEXAS SOUVENIR PLAYING CARDS
Van Noy Interstate Co., c1900. This spectacular back features a large, gold, five pointed star surrounded by cotton balls on a blue background. The faces have Type C scenes of Texas.

S63

S63 VERMONT, THE GREEN MOUNTAIN STATE SOUVENIR PLAYING CARDS
Chisolm Bros., Portland, c1910. The backs feature a photo of the State Capitol Building in a small circle with the Vermont seal in the corners. The faces show oval photo scenes (Type C).

S64

S64 WASHINGTON, THE NATIONS CAPITAL SOUVENIR PLAYING CARDS
The Waters Co., Washington, D.C., 1900. This back has a statue of the Goddess of Justice against a deep blue, starry sky with no border. The faces have oval photo scenes (Type A).

S65

S65 WASHINGTON, THE NATION'S CAPITAL SOUVENIR PLAYING CARDS
USPC, 1909. This clever back has the Washington Monument reflected in the lagoon making it reversible. The faces have oval photo scenes (Type C) of the Capital area.

S66   S67

S66 WASHINGTON, THE NATION'S CAPITAL SOUVENIR PLAYING CARDS
USPC, 1925. This back depicts the Goddess of Justice against a dark blue background framed by bas-relief columns. The faces show oval photo scenes (Type C), many of which are the same as S65.

S67 WASHINGTON VIEWS
A.C. Bosselman, New York, 1910. The backs have the same river scene as that of S39. The faces have Type D oval photo scenes.

S69

S68

S68 WASHINGTON STATE SOUVENIR PLAYING CARDS
Souvenir Card Co., Seattle & Spokane, 1899. The back has a portrait of George Washington around which are inscribed the words 'The Seal of the State of Washington 1899'. The cards come with Type A photo scenes.

S69 WASHINGTON & THE PACIFIC NORTHWEST SOUVENIR PLAYING CARDS
Lowman & Hanford, c1900. A snow-covered Mt. Rainier with fir trees and a body of water in the foreground highlights the back of this deck. The faces have oval photo scenes (Type A).

S70 WASHINGTON STATE SOUVENIR PLAYING CARDS *(not pictured)*
USPC, c1910. This issue has the same back as S69. There are new oval photo scenes (Type C) on the faces.

S71

S71 AMONG THE WHITE MOUNTAINS SOUVENIR PLAYING CARDS

Chisolm Bros., 1910. The back features the famous great stone face known as 'The Old Man of the Mountains'. The Seal of the State of New Hampshire is in each corner. The faces have Type C scenes.

S75

S75 YELLOWSTONE PARK SOUVENIR PLAYING CARDS

Haynes Photo Studios c1906. The back has a scene of the Geyser Basin in an oval frame. The oval border has the words 'Yellowstone Park'. Surrounding the oval are golden rays extending to the border of the cards. The deck comes with backgrounds of either orange or royal blue and with black or white borders, making four possible combinations. Two different sets of oval photo scenes (Type C) were used in printing this deck.

S72

S72 WHITE MOUNTAINS, NEW HAMPSHIRE, & PORTLAND, MAINE - FROM THE SEA TO THE SUMMIT

Chisolm Bros., c1900. The back has gold seals of the States of Maine and New Hampshire on a red background. Type A photo scenes were used.

S76

S76 YOSEMITE PARK SOUVENIR PLAYING CARDS

Yosemite Park and Curry Co., c1925. The back has a scene of Banner Peak in Yosemite National Park. The faces show round cornered rectangular photo scenes (type E) in the park.

S73 YELLOWSTONE PARK SOUVENIR PLAYING CARDS

F. Jay Haynes, St. Paul, c1900. The back has purple flowers and green leaves on a beige background. Printed in gold are the words 'Yellowstone Park' surrounding the monogram YNP. The faces show oval photo scenes (Type C) of features of the park.

S73

S77a

SS77 CHICAGO VIEWS

(not pictured)
A.C. Bosselman (SPCC), New York, 1900. Again the common river scene of S39 was used for the Chicago views back. The faces have oval photo scenes (Type D).

S77a CHIGAGO

SPCC, c1900. Once again, Standard produced its own version of a Bosselman deck. This deck is unusual for a souvenir deck in that it comes with a patterned back.

S74

S74 YELLOWSTONE PARK SOUVENIR PLAYING CARDS

Yellowstone Park Tourist Supply Co., c1915. This back provides a spectacular view of 'Old Faithful'. The faces have a unique style oval photo scene with a wide grey border.

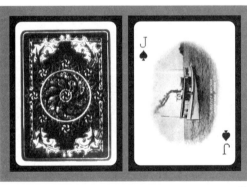

S78

S78 MACATAWA BAY SOUVENIR CARDS

1890. The maker of these cards is unknown but we know the Macatawa Bay Souvenir Co. in Michigan sold it. The deck is extremely scarce and, in fact, we know of no one who has one. It could well be the earliest souvenir deck printed in the United States.

S79

S79 CINCINNATI SOUVENIR CARDS

1909. One of the most beautiful souvenir decks ever made was issued for the Commercial Club of Cincinnati in 1909. The President at that time was John Omwake, also the President of USPC. This deck was produced in a very limited edition and came in a silk lined presentation case. It was given to guests at the annual banquet of the CCC. The courts featured officers and employees of USPC and the pip cards showed views of Cincinnati.

S80

S80 PHILADELPHIA SOUVENIR CARDS

c1960. A deck printed for and issued by the city. There is an enclosed card with a message from the Mayor. The backs feature the Seal of the City of Philadelphia. The scenes are of a new rectangular type.

S83

S83 YELLOWSTONE PARK PLAYING CARDS

Tom Jones, c1900. The back on this deck has a photograph of Yellowstone Falls. The faces of the cards have Type B scenes of features of the park.

S81

S81 VIRGINIA SOUVENIR PLAYING CARDS

USPC, 1906. This is the same deck as the Jamestown Exposition Souvenir Playing Cards with the back portraying 'The Baptism of Pocahontas'. A box has recently been discovered containing this deck with the words "Virginia Souvenir Playing Cards" on the front indicating that this deck was also sold as a Souvenir deck.

S84

S84 GREAT NORTHWEST

(Formerly SR13), Tom Jones, c1900. This deck was previously listed as a Railroad Souvenir, likely because the back pictured a train belching smoke on a single track through the forest. It has now been determined that it is not a railroad deck and is properly classified in the Souvenir chapter. The faces have Type B oval photo scenes.

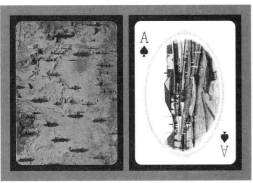

S82

S85

S82 MICHIGAN SOUVENIR

Tom Jones, c1900. This deck, with Type B photo scenes is likely by Tom Jones. Unfortunately, the pictures we have of the only known copy of this deck are quite indistinct and the details of the back cannot be determined.

S85 LANCASTER SOUVENIR PLAYING CARDS

c1900. This relatively poor quality deck was made in Lancaster, Pennsylvania, as a souvenir of the area. The back advertises the printer/publisher and the faces have Amish people and scenes from the city and surrounding countryside. It is extremely scarce with only two or three copies known.

S86 DEL CORONADO HOTEL SOUVENIR PLAYING CARDS
Alverson, Comstock Co., San Francisco, 1894. This deck features photographic scenes of the Southern California area and a picture of the hotel on the back. The faces have scenes from the area.

The last four listings are of a different type than the ones previously listed. The cards are more reminiscent of actual photographs than of playing cards and have Triplicate type indices in two corners. Some collectors believe Dougherty made these decks for Alverson, Comstock. In any event, they are very rare and command high prices in the market.

S87 LAKE TAHOE SOUVENIR PLAYING CARDS
Alverson, Comstock Co., San Francisco, c1895. The back shows the 'Tallas', a pleasure steamer taking tourists on the lake. This deck has the same Joker as S86.

S88 LAKE TAHOE SOUVENIR PLAYING CARDS
Alverson, Comstock Co., San Francisco, c1895. The back is very similar to S87 but this is a different version of the Lake Tahoe deck, with different photo scenes (some are repeated) and a 'fish' Joker.

S89 SANTA ANITA RANCH (LUCKY BALDWIN)
Alverson, Comstock Co., San Francisco, c1895. This deck, with its horse racing and gambling themes, is high on the want lists of many collectors. The back has a photo of the head of a racehorse and the words 'The Oakwood' and 'Santa Anita Ranch - Southern California'.

RAILROAD SOUVENIRS

Of all souvenir issues, probably none are more popular than the railroad decks. There are many railroad buffs throughout the world and playing cards promoting the various railway lines are a desirable item to any railroad collector. They are also considered a choice collectible to those who pursue advertising items. To the playing card collector this category is widely sought after as they are among the most interesting of all souvenir categories.

Because railroad cards cross the line into several other collecting fields they are among the most difficult souvenir decks to find. Therefore wide railroad souvenir cards, perhaps among the most prolifically issued, are often very hard to find and when found, are often expensive.

We are concerned in this chapter only with the wide railroad souvenir cards. These are the issues made for railway lines and usually sold on the trains or at the depots as souvenirs or to keep passengers occupied during their trip. The souvenir types are the decks that picture a scene along the route on the face of each card. There are probably thousands of decks with just an advertisement for the railroad line on the backs. A smaller percentage of these might have advertisements on the Aces of Spades and the Jokers. There are relatively few that fit into the souvenir category.

An excellent catalogue, The Great Book of Railroad Playing Cards, of both wide and narrow railroad playing cards was issued in 1991 by Phil Bollhagen.

The first six listings are all by Interstate News Co., a supplier of goods to the stations and trains. They likely all date from around 1900 to 1910 and were, no doubt, manufactured by USPC. Each of these six decks has oval photos (Type C) showing various scenes along the routes.

SR1 CHICAGO, MILWAUKEE & ST. PAUL RAILWAY SOUVENIR PLAYING CARDS
The back features a scene of the Falls of Minnehaha behind a rustic footbridge crossing the stream. In a panel below, there is a complete stanza from Longfellow's 'Hiawatha' that starts "Where the falls of Minnehaha flash and gleam among the oak trees".

SR1

SR3 ALONG THE C.M. & ST. PAUL - LAKE MICHIGAN TO PUGET SOUND
The back of this deck has a round photo on an orange background of a two-unit box cab electric locomotive on a single track. 'Electric' is printed in red above and reversed below. The print is surrounded by gold shock waves.

SR3

SR2 CHICAGO, MILWAUKEE & ST. PAUL RAILWAY SOUVENIR PLAYING CARDS
The back pictures a scene of snow capped mountains in the background with a grove of tall fir trees in front, all reflected in a mirror-like lake.

SR2

SR4 ALONG THE C.M. & ST. PAUL - LAKE MICHIGAN TO PUGET SOUND
This back is similar to SR3 with the photo in the center depicting a bi-polar electric engine pulling a passenger train along a double track.

SR4

SR5

***SR5 ALONG THE C.M. &
ST. PAUL - LAKE MICHIGAN
TO PUGET SOUND***
*The back is again similar to the
previous two listings with this photo
featuring a bi-polar electric engine
pulling a long passenger train along
the center of three tracks. The faces
have the same oval photo scenes as
SR4.*

SR10

***SR10 DENVER & RIO GRANDE,
WESTERN & WESTERN
PACIFIC RAILWAYS
SOUVENIR PLAYING CARDS***
*Van Noy Interstate Co., c1915. The
back has a scene of a train passing
through a river gorge along the route.
State Seals of Colorado, Utah, Nevada,
and California are in the corners. The
faces have rectangular photo scenes
(Type E).*

SR6

***SR6 ALONG THE C.M. &
ST. PAUL - LAKE MICHIGAN
TO PUGET SOUND***
*This back is similar to SR2 with a
snow-capped mountain in the
background with the entire scene
outlined by fir branches.*

SR11

***SR11 DENVER & RIO GRANDE
R.R. SOUVENIR PLAYING CARDS***
*Tom Jones for Van Noy Interstate Co.,
Denver, c1900. The back shows a
cloudy background in either dark blue
or green. In the semi-circle surrounding
a small scenic sketch are the words
'Denver & Rio Grand RR' and below
the scene the words 'Scenic Line of the
World'. The faces have the oval photo
scenes (Type B) typical of Tom Jones
decks.*

SR7   SR8

SR14

SR12

SR7 C&O RAILWAY SOUVENIR PLAYING CARDS
*USPC, c1908. The back has a brown and orange filigree frame
surrounding a line sketch of the front of a locomotive. In the center of
the locomotive are the letters C & O with 'Safety First' printed above
and below. The faces have oval photo scenes (Type C).*

SR8 C&O RAILWAY SOUVENIR PLAYING CARDS
*USPC, c1900. The back is yellow with a brown design. The words
'C&O' are in a panel at the top and on the bottom panel are the letters
'FFV'. The center strip reads 'Via Washington' with 'East' and 'West'
above and below. The faces have Type A oval photo scenes.*

SR9 C&O RAILWAY SOUVENIR PLAYING CARDS *(not pictured)*
*USPC, c1905. The back is identical to SR8 but has an orange
background. The Type C scenes in this issue are different than SR8 but
the same as some editions of SR7. There are many known variations of
the scenes and individual cards may differ in both decks.*

**SR12 DENVER & SALT LAKE R.R. MOFFAT LINE SOUVENIR
PLAYING CARDS**
*USPC for Barkalow Bros., 1914. The back features a woman on a cliff
waving her handkerchief with a river running far below. The faces
have Type C oval photo scenes.*

SR13 GREAT NORTHWEST *(not pictured)*
This listing has been reclassified as S84.

**SR14 GREAT NORTHERN PACIFIC STEAMSHIP CO., -
SPOKANE, PORTLAND & SEATTLE RAILWAY CO.,
OREGON ELECTRIC RY CO., OREGON TRUNK RAILWAY
SOUVENIR PLAYING CARDS**
*USPC, c1915. The back has a large oval framing a black and white
steamship with two yellow funnels. There is a steamship company seal
in each of the four corners. Great Northern Pacific Steamship operated
five different railroad lines. The faces have the later rectangular photo
scenes (Type E).*

SR15 SOUVENIR PLAYING CARDS OF THE GOLDEN WEST
Van Noy Interstate Co., c1910. The back has a lake with two large mountains in the background. There is a man in a rowboat and two horses on the shore. The faces have oval photo scenes (Type C).

SR15

SR16 SOUVENIR PLAYING CARDS OF THE GOLDEN WEST
Van Noy Interstate Co., c1910. The back features a snowy Mt. Shasta and a village with a large fir tree in the foreground. The faces have the same oval photo scenes as SR15 (Type C).

SR16

SR17   SR18

SR17 SOUTHERN PACIFIC SOUVENIR PLAYING CARDS
Barkalow Bros., Denver, c1900. The back shows either a sepia or grey toned photo of a train crossing the Great Salt Lake on a two-track trestle on the left side of the card. The sunset is reflected in the water and in small white print is the title 'Evening on Great Salt Lake'. The faces have Type B oval photo scenes.

SR18 SOUTHERN PACIFIC LINES SOUVENIR PLAYING CARDS
Barkalow Bros., Denver, c1915. The back has the same scene as above but it is in full color and there is no title. The faces have Type C oval photo scenes that are different than SR17.

SR19 SOUTHERN PACIFIC LINES SOUVENIR PLAYING CARDS
Barkalow Bros., Denver, c1925. This back shows a yellow train moving forward on a trestle over a green body of water with the sun setting in the background. The faces have the same SR18 oval photo scenes.

SR19

SR20 SOUTHERN PACIFIC LINES SOUVENIR PLAYING CARDS
USPC, c1930. The back features a red, orange and black train moving to the right with purple mountains in the background. On the left are palm trees and a tower. The faces also have the same oval photo scenes as SR18.

SR20

SR21   SR22

SR21 SOUTHERN PACIFIC LINES SOUVENIR PLAYING CARDS
USPC, c1935. This is the last Southern Pacific wide issue. The back shows a train crossing a bridge with men in a cotton field. The train can be red/yellow or red/grey and the cotton fields either green or brown. There are therefore four possible varieties. The faces have Type E rectangular photos.

SR22 OVERLAND ROUTE SOUVENIR PLAYING CARDS
Barkalow Bros., Omaha, c1910. The back features a high green mountain with the Devil's Slide down the center of the picture. The 'Union Pacific' Shield is at the top and 'Devil's Slide, Utah' is printed in gold beneath the picture. The faces have oval photo scenes (Type C).

SR23

SR24

SR23 UNION PACIFIC PICTORIAL SOUVENIR PLAYING CARDS
Barkalow Bros., Omaha, c1930. The back shows two riders on a trail overlooking the harbor. Printed at the lower right is "Scenic Trails overlook Avalon Bay at Santa Catalina Island, California". The faces have oval photo scenes (Type C).

SR24 WESTERN PACIFIC SOUVENIR PLAYING CARDS
USPC for Van Noy Interstate Co., c1915. The back features a deep gorge with a bridge spanning a river far below. The faces have rectangular photo scenes (Type E).

SR25 ROCK ISLAND SOUVENIR PLAYING CARDS
USPC, 1910. The back has black printing on an orange background with a large globe in the center and a train, with the Rock Island shield, above and beneath it. On both sides of the card are the words 'Best Service West'. The faces have oval photo scenes of Type C.

SR25

SR29 WHITE PASS & YUKON SOUVENIR PLAYING CARDS
USPC, c1900. The backs are the same as SR28 except that above and below the oval in white lettering on dark blue are the words 'Scenic Railway of the World'. The faces have the same Type A oval photo scenes.

SR29

SR26 ROCK ISLAND SOUVENIR PLAYING CARDS
John J. Grier News Service, c1910. The back has a train moving to the left on double tracks with blue sky and clouds overhead. The faces have Type C oval photo scenes.

SR26

SR30 WHITE PASS & YUKON SOUVENIR PLAYING CARDS
USPC, c1900. The background of this issue has a similar design to the previous two listings. This deck comes with a blue or red back. There is a large circle in the center of the card that is either blue (on the red background) or red (on the blue background). Large white letters in the circle read 'White Pass and Yukon Route' and surround a small blue sketch of a train. At the bottom of the sketch are the words 'Gateway to the Yukon'. The faces have different Type A oval photo scenes.

SR30

SR27 SEABOARD RAILWAY SOUVENIR PLAYING CARDS
USPC, c1910. The back shows orange blossoms on a tan background with a circle in the center which has 'Seaboard Airline Railway' printed on it. Inside this is a red heart containing the words 'Through the Heart of the South'. The faces have oval photo scenes (Type C).

SR27

SR31 WHITE PASS & YUKON ROUTE SOUVENIR PLAYING CARDS
USPC, c1910. The backs have a patterned background with a circle in the center that is either blue or red. The train in the sketch does not reach into the outer circle like SR30. The oval photo scenes in this deck are Type C and different than the other three listings.

SR31

SR28 WHITE PASS & YUKON SOUVENIR PLAYING CARDS
USPC, 1900. The back has a red and white background featuring a large white oval outlined in gold. Within the oval is a blue sketch of a railroad scene. Printed in red are the words 'The White Pass and Yukon Route'. The faces have oval photo scenes (Type A).

SR28

SR32 THE SOO LINE
Tom Jones (USPC), Cincinnati, c1895. This wide souvenir for The Soo Line is very scarce with a picture of a craggy mountain in an oval surrounded by red maple leaves, reminiscent of certain Canadian souvenirs. The deck has Type B photo scenes. The Joker is an ad for The Soo Line, not a souvenir photo.

SR32

CANADIAN AND OTHER SOUVENIR ISSUES

Canadian souvenir decks were often made in England with later issues from both Canada and the United States. In style they are very similar to the U.S. types with photo scenes on each card and backs that portray some aspect of the subject of the souvenir deck. Certain Canadian souvenir decks have new types of photo scenes, classified as Types F and G. These photo scenes feature a mixture of oval and rectangular scenes, some of which extend beyond the borders. The decks listed here include all the wide province and city souvenirs, as well as railroad souvenir issues.

SCA5

SCA5 ATLANTIC OCEAN TO THE HEAD OF THE GREAT LAKES
Goodall, c1905. This deck features yet another beauty with a green dress and brown hat and the name Grand Trunk Railway System in the upper right hand corner. The photos (Type F) show scenes along the railroad route.

SCA1   SCA3

SCA1 BRITISH COLUMBIA SOUVENIR
Clarke & Stuart, Publishers, Vancouver, BC, c1895. The back of this deck made by USPC features a lake surrounded by maple leaves in vibrant colors. The faces show oval photo scenes (Type A) of the area.

SCA2 BRITISH COLUMBIA SOUVENIR (not pictured)
Clarke & Stuart, Publishers, Vancouver, c1900. Another B.C. souvenir made by USPC with the same back but different photo scenes (Type C) on a number of the cards.

SCA3 ATLANTIC OCEAN TO HEAD OF THE GREAT LAKES
Goodall, London, c1905. This back has a navy blue border and features a blond woman wearing a pink dress and bonnet surrounded by deep pink flowers. The photos, Type F (similar to Type D but with smaller photo scenes), picture scenes along the Grand Trunk Railway System.

SCA6

SCA6 GRAND TRUNK PACIFIC RAILWAY
Goodall, c1905. A still different woman surrounded by a navy blue border featuring the name Grand Trunk Pacific Railway in the corners. The photo scenes are again Type F.

SCA7

SCA7 MONTREAL & QUEBEC SOUVENIR
Canadian Playing Card Co. Ltd., Montreal, c1920. This deck features a crest with French fleur-de-lis, maple leaves and the English crown and lion on a yellow background. The photos show scenes of the two cities and are of a type similar to Type D but with some scenes protruding through the borders. We will describe these as Type G.

SCA4

SCA4 ATLANTIC OCEAN TO THE HEAD OF THE GREAT LAKES
Goodall, London, c1906. This beautiful back features a woman wearing a green dress with a yellow hat surrounded by pink and blue flowers. It comes with either a green or purple border. These cards, also for the Grand Trunk Railway System, have different photo scenes (Type F) than SCA3.

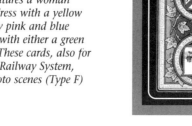

SCA8

SCA8 MONTREAL & QUEBEC SOUVENIR
Goodall, c1905. This back features double ended circles, each enclosing a different crest and surrounded by maple leaves. The back was published in four very different color combinations with dark and light borders. These decks were published with several different sets of Type F photo scenes.

SCA9   SCA10

SCA14   SCA15

SCA9 MONTREAL & QUEBEC SOUVENIR
Goodall, c1907. This deck, with Type F scenes, also features double ended crests but not in circles.

SCA10 MONTREAL & QUEBEC SOUVENIR
Goodall, c1905. The colorful back of this deck again features the crests although they are somewhat larger and more ornate. The photo scenes (Type F) seem slightly earlier although depicting the same scenes as SCA9.

SCA14 OCEAN TO OCEAN
Goodall, c1905. This back is very similar to SCA13 but the bottom of the crest reads "Ocean to Ocean". The same oval photo scenes were used.

SCA15 OCEAN TO OCEAN
Goodall, c1900. This deck features the nine seals of the Canadian provinces and territories at that time: Ontario, Quebec, Nova Scotia, Prince Edward Island, New Brunswick, Manitoba, British Columbia, Yukon and the Northwest Territories. The deck has a white or pale blue border and the earlier editions come with the standard Goodall Ace of Spades. The photos show scenes of Canada (Type G).

SCA11   SCA12

SCA16   SCA17

SCA11 OCEAN TO OCEAN
Goodall, c1912. This deck honors King George V who succeeded to the throne of England in 1911. There are two different color backs featuring the King and the eleven crests of the provinces and territories at that time. The photos are of scenes across Canada (Type G).

SCA12 OCEAN TO OCEAN
Goodall, c1915. A slightly later issue, which also was available in two colors, with the same crests and featuring the King wearing a plumed hat. The scenes are of Canada, coast to coast (Type F).

SCA16 OCEAN TO OCEAN
Goodall, c1905. Two additional provincial seals were added to the backs of this deck, Saskatchewan and Alberta, and the deck can be found with a light blue or light green background. The cards have oval photo scenes (Type G).

SCA17 PICTURESQUE CANADA
Goodall, c1910. This deck was published for the Canadian Pacific Railway. The cards show scenic views (Type F) of Canada and the Joker features Hiawatha, an Ojibway.

SCA13

SCA13 OCEAN TO OCEAN
Goodall, c1905. This back features a large crest and reads 'Canada' at the bottom. The photo scenes are views across Canada (Type G).

SCA18

SCA18 PICTURESQUE CANADA
Canadian Playing Card Co., Montreal, c1920. This deck was published by Valentine & Sons. The backs feature the Valley of Ten Peaks, Moraine Lake. The scenes are Type G.

SCA19   SCA20

SCA24   SCA25

SCA19 PICTURESQUE CANADA
Goodall, c1910. This deck was published for the Canadian Pacific Railway News, a company that sold food, newspapers and souvenirs on CP trains. The back features a lake scene with pine trees in the foreground in a square frame with maple leaves. The scenic views (Type G) were taken along the railroad line.

SCA20 PICTURESQUE CANADA
USPC, Windsor, Ontario, c1920. Another deck done for the Canadian Pacific Railway News. The back features a view of the Banff Springs Hotel with either a light green or light yellow border. The faces had oval photo scenes (Type G), but in a larger format.

SCA24 TORONTO SOUVENIR
Goodall, c1915. The next four decks, all by Goodall, feature a different seal in the center of each back. The known issues (would they not have made them for more provinces/cities?) are for the City of Toronto, British Columbia, Manitoba and Ontario. The faces have oval photo scenes (Type F).

SCA25 BRITISH COLUMBIA

SCA21   SCA22

SCA26   SCA27

SCA21 PICTURESQUE CANADA
Consolidated Lithographing Ltd., Montreal, c1912. Made for the Canadian Pacific Railway News Service, this deck features the same back as SCA20 but with a gold border. It has different oval photo scenes (Type G).

SCA22 PICTURESQUE CANADA
Consolidated Lithographing Ltd., Montreal, c1912. This was the mate to SCA21 with the back featuring the Chateau Frontenac Hotel in Quebec City and using the same set of oval photo scenes.

SCA26 MANITOBA   SCA27 ONTARIO

SCA23 PICTURESQUE CANADA,
USPC, c1900. This deck is an earlier issue, for Canadian Pacific Railway, with the backs featuring an Indian at the foot of the Rocky Mountains. The Type C oval photo scenes are of Canada and Alaska.

SCA23

SCA28

SCA28 INTERCOLONIAL RAILWAY
Goodall, c1905. There are three different back designs for the Intercolonial Railway. The pictured Ace of Spades was not scenic and each back featured a moose head within a circle. The Joker (also not scenic) gave pertinent facts about the railway system. Each deck had identical photo scenes (Type G).

SCA29   SCA30

SCA29 INTERCOLONIAL RAILWAY

SCA30 INTERCOLONIAL RAILWAY

SCA31

SCA32

SCA31 WINNIPEG SOUVENIR
Published by Wm. A. Martel & Sons, Winnipeg, Manitoba, c1895. The backs on this deck are red and white and feature a buffalo, beavers and wheat. The faces have scenic photo scenes (Type D).

SCA32 NEWFOUNDLAND SOUVENIR
USPC, Cincinnati, published by Ayre & Sons Ltd., St. Johns, Newfoundland, c1920. This deck features the great Newfoundland dog on the back and comes in both red and blue. The scenes are rectangular photos (Type E).

SCA33 OCEAN TO OCEAN
For Canadian Pacific Railway, Goodall, c1910. The back on this deck has scrolling and encloses an oval on tan picturing a snow-covered mountain overlooking a valley of green trees. There are circles in each of the corners surrounding a maple leaf. The photo scenes are of Canada from Quebec westward, with many mountain views.

SCA33

OTHER SOUVENIRS

The following decks, with the exception of SCU3 and SO3, were made by USPC and are thus included here. No doubt there were more of these souvenir decks produced for other countries.

SCU1   SCU2

SCU1 CUBA
By USPC for H.T. Roberts, Havana, c1915. The backs are in three different combinations, black and white, black and blue, and black and yellow. The faces have rectangular photo scenes (Type E).

SCU2 CUBA
For Romo & Kredi, c1905. The backs are gold on black and feature dancers. The faces are oval photo scenes (Type C).

SCU3   SO1

SCU3 CUBA
Unknown Cuban manufacturer, c1930. Although these cards look older and cruder, they are probably the latest published. The backs are red, blue and white and signify an alliance between the USA and Cuba. They are square cornered, smaller, and with Type C scenic photos.

SO1 SOUTH AFRICA
USPC, c1910. A fine deck made for sale in South Africa and featuring scenes taken throughout countries in that region.

SO2   SO3

SO2 TASMANIA
USPC, c1908. Another souvenir deck made for export and featuring scenes from the area.

SO3 PERU
Standard PCC, c1910. A souvenir issue made for sale in Peru.

COLLEGES, UNIVERSITIES AND UNIONS

The earliest known United States university playing cards appear to have been ones designed for Yale, Harvard and Princeton Universities and issued in 1896. In this chapter we have listed the known decks with special backs, Aces of Spades and Jokers as well as some general numbers for newer decks.

While we are unaware of other decks published by Marshall or Ward's, it is quite possible that others promoting different universities will show up.

CU2

CU2 PRINCETON UNIVERSITY
L.W. & H.N. Marshall, 1896. This deck is similar to the Yale deck but made for Princeton. It has the same type of Ace of Spades, a more decorative back with the Princeton tiger and, presumably, a special Joker.

CU1

CU1 YALE UNIVERSITY
L.W. Marshall, 1896. This is a beautiful deck, with an Ace of Spades and Joker intricately designed. Both of these cards have sketches of the campus areas and buildings. The remainder of the deck is standard. This deck has a special back design with the Yale pennant prominent.

CU3

CU3 IVY LEAGUE PLAYING CARDS - YALE
Ward's, Boston, Mass., c1900. These cards come in a handsome slip case. Like the above two decks, the college interest is confined to the Ace of Spades, Joker and backs. Ward's of Boston was a stationer and the two known examples were done for Yale and Harvard.

CU1a

CU1a HARVARD UNIVERSITY
L.W. Marshall, 1896. Collectors had often wondered why Marshall had not made a deck for Harvard. Recently a Harvard version was found with the interesting Ace proudly repeating the Harvard cheer. Unfortunately the Joker has yet to be seen.

CU3a

CU3a IVY LEAGUE PLAYING CARDS - HARVARD
Ward's, Boston, Mass., c1900. The Harvard version is very similar to the Yale cards. Both decks have backs that match the pictured Jokers.

273

CU4

CU6

CU6 COLLEGE CARDS
c1925. These cards have back interest only. There were literally thousands of different backs issued over the years as most institutions would have had playing cards printed for their own use at one or more points in their history.

CU4 CONGRESS #606
USPC, 1901 to 1919. Congress playing cards were known for their picturesque backs. Six designs were done for universities – Michigan, Harvard, Yale, Princeton, Cornell and Pennsylvania. All of the backs had designs in their school colors with pennants surrounding mottoes, campus pictures, mascots and other pertinent subjects in the center of the card. While likely popular, as they were issued for several years, they are nevertheless quite scarce and in high demand by collectors

CU7 COLLEGE ATHELETES *(not pictured)*
c1970. Another listing category, this time for universities and colleges honoring their athletic stars with pictures on playing cards made to be sold in bookstores and shops near the institution.

CU5

CU5 COLLEGE CARDS
c1920. This listing is for other college and university playing cards which have special Aces of Spades, backs and, sometimes, special Jokers. They were mostly issued after WWI and were usually narrow. A few examples from the period between 1920 and 1935 are shown above.

A number of interesting decks issued by labor unions and companies do not promote 'products' and thus do not fall into a specific advertising category. The special ones are categorized in this chapter (note the change from 'U' to 'Un' in the listings to avoid confusion with the early makers which also used 'U' in the original Encyclopedia). Union decks which only have back interest remain in the ADD category.

A. Eldon Duke issued the earliest deck of the special union type in 1910. Mr. Duke was a member of the International Typographical Union and issued a deck of cards to promote Labor's message.

Un1

Un3

Un1 ELDON DUKE
1910. Every card in this deck either carries a facsimile of a union label or has a union message. Besides being a regular deck it was designed to play the "Great Union Card Game" described in an accompanying book of rules.

Un3 THE NATIONAL LITHOGRAPHERS DECK
USPC, 1915. A souvenir type of deck published by the National Lithographer's Society where each card has a either a photograph of an officer of the society, a picture of a important contributor in the field of lithography, a lithographic process in the 1915 era or a historic event in the field.

Un2

Un4

Un2 UNITED DRUG COMPANY
Boston, Mass., 1913. This deck was a souvenir of the 11th annual convention of the United Drug Company in Boston in 1913. The Company manufactured pharmaceuticals and operated a chain of retail drug stores. Each card in the deck has a photograph of an officer in United Drug or in one of the various state Rexall clubs.

Un4 THE UNITED BROTHERHOOD OF CARPENTERS AND JOINERS
1920. This is a standard deck with a special Ace of Spades and Joker. The backs and the box also mention the union.

Un4a

Un4a CIGAR MAKERS INTERNATIONAL UNION
c1912.There were probably other decks of this type, issued by different unions. Decks with these distinctions should be given this listing number.

Un5

Un5 INTERNATIONAL PRINTING PRESSMAN AND ASSISTANTS UNION
International Playing Card Co., Chicago, IL. This beautiful round deck was a special printing of 016 for this union.

Un6

Un6 AMALGAMATED MEAT CUTTERS
Brown & Bigelow, c1952. This deck was issued for the Amalgamated Meat Cutters and Butcher Workmen of North America. Each card has a picture of a pioneer of the labor movement in the United States, their accomplishments and their birth and death dates. One of the Jokers reads "In Unity there is Strength", and the other, "Divided We Fall". The backs picture their National Headquarters in Chicago.

BRIDGE AND WHIST

The playing cards featured in this chapter are those that were specifically designed to teach Whist as well as either Auction or Contract Bridge. We also feature decks that were used for playing duplicated hands in competition, International or otherwise. Duplicate Whist and Bridge were invented in an effort to reduce the luck of the deal in competitive tournament events by allowing all competitors to play the same hands throughout the course of a tournament session.

Bridge is the only card game that continues to have annual International competitions. It is also the only game where four people, each one from a different part of the world and each speaking a different language, can join together and play a serious card game. Bridge transcends all boundaries and it is really the only game where special decks were designed as an aid to help players in the play of the cards.

BW2

BW2 FOSTER'S SELF-PLAYING WHIST CARDS
First Series, Andrew Dougherty, New York, 1889. An ingenious method of Whist instruction that can be used by 1 to 4 players at a time. The backs of the cards are covered with numbers. The first number at the top left corner is the number of the deal (1 to 40). The letter beneath that is the location of the card; A is dealt to the dealer's left, B to the right, Y to his partner and Z to himself. The number beneath the letter is the order in which that card should be played (13 to 1). Each deal is analyzed in an accompanying booklet. When it is a player's turn to play, he selects the card he thinks should be played. If correct, it will match the appropriate number on the back. If only 1, 2, or 3 players are participating, they can sort the cards for the missing player(s) according to the backs, and turn them up during the trick. Finally, an extra card tells what trump card was turned up.

BW1

BW1 AMES WHIST LESSON CARDS
Russell & Morgan, Cincinnati, for F. Ames, 1889. This is the earliest of the cards that teach during play. Whist, the forerunner of Bridge was played with no bidding and with all hands concealed (no dummy). Trump was determined when the dealer turned the last card face up. In the Ames Whist Cards every honor card (A, K, Q, J and 10) provides suggestions as to when that card should be played.

BW3 FOSTER'S SELF-PLAYING WHIST CARDS *(not pictured)*
2nd Series, Andrew Dougherty, New York, 1891. The first series advertised 160 hands, really better described as 40 deals. The second series was reduced to 128 hands or 32 deals.

BW4 FOSTER'S SELF-PLAYING BRIDGE CARDS
NYCC, New York, 1903. When the birth of Auction Bridge started the demise of Whist, R.F. Foster applied his patent to Bridge. The same principles were used but the hands were different and the bidding was included in the descriptive material.

BW4

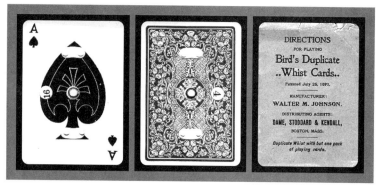

BW5

BW5 BIRD'S DUPLICATE WHIST CARDS
Walter M. Johnson, Somerville, Mass., 1893. The first special deck that was engineered so that the same deal could be replayed numerous times. The numbers on the face of the cards indicate the deal number (1 to 16). The numbers on the backs of the cards indicate its dealing location (1 to 4). A smaller number on the back indicates the trump card and this card is turned to indicate trump, as in regular Whist. When the deal is finished, all the cards are stacked and, with the aid of a special needle inserted in the left double hole and moved to the right, each card is advanced to the next deal. When 16 deals are completed the set is then passed to the next table where it may be played again to provide a duplicate competition.

BW6

BW6 VIRGINIA BRIDGE CARDS
Ruledge Playing Card Co., Port Huron, Mich., 1908, copyrighted by Virginia M. Meyer. This is another learn-as-you-play deck. Every card has Bridge rules along the margins. Many of these rules were carried over from Whist and make little or no sense to the modern Bridge player.

BW6a

BW6a VIRGINIA SKAT CARDS
Ruledge Playing Card Co., Port Huron, MI, 1908, copyrighted by Virginia Meyer. Another deck using the same general format as above and the only one known for this game.

BW7

BW7 AMERICAN WHIST LEAGUE PLAYING CARDS
Andrew Dougherty & Co., New York, c1890. The AWL had special cards made for them to use in their tournaments and to sell to their affiliated clubs. The introduction of indices on playing cards was a great advantage to the game of Whist, as it allowed a player to hold 13 cards in his hand and see all of them. This allowed players to play in a more natural manner.

BW8

BW8 AMERICAN WHIST LEAGUE PLAYING CARDS
National Playing Card Co., Indianapolis, Indiana, c1895. The cards in this AWL deck were classified as Grade B.

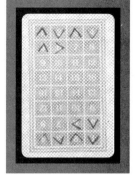

BW8a

BW11

BW8a AMERICAN WHIST LEAGUE PLAYING CARDS
American Card Co., Kalamazoo, Michigan. These decks were probably sold to the league already packaged for play as the known examples are in a paper folding duplicate whist tray.

BW11 EDUCATOR BRIDGE
USPC, c1925. This deck was issued with a numbered back to allow a teacher to prepare lesson hands. A duplicate bridge tournament could also be held by marking the backs the first time the hand is played.

BW9

BW9 PAR AUCTION
By Milton C. Work, Milton Bradley Co., Springfield, Mass., c1920. R.F. Foster was considered the top American authority on Whist and this distinction was awarded to Work on Auction Bridge. There are many different decks in this Par Auction series. The idea was copied from Foster's decks but with a simplified back. Around each number on the back is an arrow which gives the direction that specific card should be dealt for that deal. In this series, each deal has a par score that should be attained if played correctly. When the set is completed the booklet, which accompanied the pack, can be referred to determine what should have been done and why.

BW12

BW12 POINT COUNT DOTS
New York Consolidated Card Co. (Samuel Hart), c1915. Milton Work devised a system of bidding in Auction Bridge based on point count, Aces being assigned 4, Kings 3, Queens 2 and Jacks 1. To aid the new bridge player in evaluating his hand he had these cards designed with the appropriate amount of dots next to the index. This deck was made in a smaller size to make it easier for the card player to hold all 13 cards.

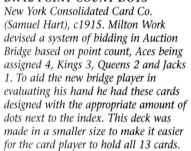

BW13

BW13 GEM #51
New York Consolidated Card Co., c1915. These decks were also made in a smaller size and with the point count dots. They measure 2 1/4 x 3 inches.

BW10

BW10 LESSONS IN AUCTION
By Milton C. Work, Milton Bradley Co., c1920. This is also one deck in a large series. Whereas Par Auction was designed for the more advanced players, Lessons in Auction was meant for beginners. It had 16 deals per deck.

In 1926, Harold S. Vanderbilt and some of his associates invented Contract Bridge. During the years 1926 to 1929 it slowly replaced Auction Bridge. This change accelerated in 1929 when Ely Culbertson, who was one of the greatest promoters of all time, determined to become 'the authority' on Contract Bridge.

In the early days of Contract Bridge, there were three popular bidding systems, Culbertson's Approach Forcing, Sims' One Over One and the Official System, which was devised by a group that included Milton Work. Almost all the teaching decks of the early Contract days were based on one of these systems.

BW14

BW14 CREST APPROACH FORCING DECK
Crest Card Co., Evanston, Illinois, c1932. This deck is based on Culbertson's system with all of the aces and court cards having bidding aids in the margins.

BW15

BW15 CREST OFFICIAL SYSTEM DECK
Crest Playing Card Co., c1932. This is essentially the same deck as BW14, except it has completely different markings in the margins of the aces and courts.

BW18

BW18 EASIBID
E.E. Fairchild Co., Rochester, NY, 1931. This deck was adapted for the Approach Forcing system by Culbertson. Margin instructions are on aces, kings and queens only. A complete instruction card came with each pack that outlined the opening bids. The deck's motto is "A Bid in the Hand is Worth Two in the Pack".

BW19 EASIBID *(not pictured)*
NYCC, New York, 1931. This Easibid deck was adapted for the Official System. It was made on the 'Deluxe' brand bridge size cards with a different system, but it shared the same Registered Trademark as BW18.

BW17

BW17 HUNTLEY OFFICIAL SYSTEM DECK
Bid-Rite (Brown and Bigelow), 1932. A similar deck to the above used the Official System. Both decks contain a booklet that briefly outlines the systems.

BW19a

BW19a EASIBID
USPC, New York, 1931. Another version was produced by USPC in this very competitive era.

BW20

BW20 CULBERTSON'S OWN PLAYING CARDS
Russell Playing Card Co., New York, 1932. A deck designed by Ely Culbertson promoting the Approach Forcing System. All 52 cards feature bid and play instructions in the margins. A booklet outlining the system came with the pack.

BW16

BW16 HUNTLEY APPROACH FORCING DECK
Bid-Rite (Brown and Bigelow), 1932 (also NR12). This four color, no revoke deck has centered boxes containing bidding and play information on all 52 cards. It was based on Culbertson's Approach Forcing System.

BW21

BW21 UTILITY DECK
*Utility Card Co., New York, 1933. This series goes back to the concept
of prearranged decks for Duplicate Bridge. Each deck had 20 deals and
a 140 page booklet explains each deal and provides commentary on the
bidding. A second Ace is also shown which has a slightly different back.*

BW22

BW22 PREARRANGED DUPLICATE CARDS
*James Bell Co., Newark, 1935. Little is
known of this series as there is only a
sample card available.*

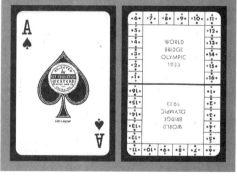

BW23

BW23 1933 WORLD BRIDGE OLYMPIC
*Western Playing Card Co., Racine, Wis. This deck was made for the
National Bridge Association, New York, with hands selected by Ely
Culbertson. There were 16 hands that were played at the same time
all over the world. Directors of these games were supplied with the
decks, and none were to be opened before 8:01 PM, May 1, 1933.
Around each number on the backs of the deck are four dots - one red,
one green, one gold, and one purple. Each contestant was assigned a
color and played the hand assigned to that color. Of the 16 prepared
hands, each color had four par problems. The players played each
deal with different partners, and against different opponents. The
highest score around the world was declared the winner and fancy
trophies were awarded for overall and local winners.*

BW24 1934 WORLD BRIDGE OLYMPIC (not pictured)
*Western Playing Card Co. Racine, Wis. The same deck as above, but
with new hands for the following year's contest.*

About 1944 Charles Goren, who brought Contract
Bridge to millions of bridge players, picked up the torch
that had been carried by Foster, Work, and Culbertson.
He simplified the bidding by creating a system out of the
basic point count theory originated by Work 25 years
before. He added distributional points and many
refinements that made Contract Bridge easier to play for
people who had previously found it too complicated.

BW25

BW25 BRIDGEPOINT PLAYING CARDS
*The Bridgepoint Playing Card Co., Philadelphia, Penn., 1945. These
were standard decks, with a simple 1, 2, 3, or 4 under the indices of
the appropriate card.*

BW26

BW26 POINT COUNT BIDDING
*Brown & Bigelow, c1945. This deck was
issued with the same idea as BW25, but
the point count numbers were in blue
and somewhat larger.*

BW27

BW27 IDEAL BRIDGE
*Atlantic Playing Card Co., New York, for the Bridge Hand of the Month
Club, Bayside, N. Y., 1955. This deck is a modern version of Bird's
Duplicate Whist (BW5). There are 20 prepared hands in each deck. When
aligned, and the needle inserted as per the instructions, the hands are dealt.
After completion of the deals a booklet analyzes each hand and gives the
correct play. There were two decks to each set with a new set to be published
monthly.*

BW28

BW31

BW28 PLAY BRIDGE WITH GOREN
Milton Bradley, Springfield, Mass, 1967. There were several editions of this deck. Each one pictured Goren on the Joker and contained a point count bidding card. There was one edition called Bridge for One. This is a solitaire game in which hands can be bid and played. In this edition each suit had the suit signs also printed on the backs. There was also a Bridge for Two set.

BW31 MILTON WORK BRIDGE SET
USPC, 1928. A double deck set, each with 52 cards and "valuable and dependable information" on extra cards.

BW29

BW29 AUCTION BRIDGE
J. L. & B. A. Rogers, Kansas City, Missouri,1930. This deck has 52 cards plus four instructional cards and a scorecard. There is an unusual seal on the diamond two, perhaps a registry of a prototype.

BW32

BW32 TWELVE TELLING TESTS
Bridge Headquarters Inc., New York, 1934 Series. A deck of 52 cards with dealing backs and a book on bidding and play.

BW30

BW30 MASTER BRIDGE DEAL CARDS
MB Institute Inc., New York, 1931. This set includes a dealing tray, special dealing frame and pinhole dealing cards, a deck of 52 cards and three sets of deals. The deals are taken from tournament play including the 1931 Vanderbilt Cup. Madeline Kerwin and Edward C. Wolfe selected the hands.

BW33

BW33 BRIDGE HEADQUARTERS
USPC, 1933. This is one of a number of bridge decks published by Bridge Headquarters which feature special aces.

BW34

BW35

BW34 WHIST CARDS FOR PRACTICE

S. T. Varian, East Orange, N.J., 1893. These have small cards as a border around the faces of each card. The small cards tell the leader to a trick what card should be played in either the trump suit or a plain suit, from various combinations. The decks could be had with either standard or American style leads.

BW35 CULBERTSON
Kem Card Co., c1931. A standard Kem deck with an interesting Josephine Culbertson extra card given as a prize in a contest sponsored by Redbook Magazine.

BW36

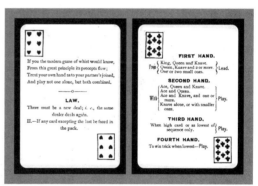

BW37

BW36 GEORGE GOODEN
c.1940. A later version of a special dealing back.

BW37 WHIST INSTRUCTION
c1880. An early instructional deck, which based on the Triplicates and back design, we surmise was made by Dougherty.

Since World War II a number of newer Bridge teaching and dealing decks have been produced to help bridge players improve and to allow competition. We have not attempted to cover these in this work.

MacLean's Magazine, November 15, 1929

1929 AD

NO-REVOKE CARDS

For centuries even the most expert card players have occasionally played the wrong card in error. In Bridge, Whist, or similar games, when a player fails to follow suit when he is able, he has revoked. In many cases a card gets slotted in one's hand by error with the other suit of the same color, i.e. a heart in with the diamonds, or a spade in with the clubs. For years decks have been designed in an attempt to help prevent these mistakes. Introducing decks with four different colored suits has been the most common method, but there have been others to try to make card reading more foolproof for the average card player. It is surprising that none of these decks which proved to be effective are being used today. Finally, it should be noted that many four-color suit decks were not issued as no-revoke decks. Although some are listed here, others, such as the round decks by Globe and Richardson, are only listed in the appropriate chapter.

NR1

NR1 THE SEMINOLE WARS DECK
J.Y. Humphreys, Philadelphia, c1819 (W2). This is a beautiful, hand-colored deck with Presidents representing the kings, Goddesses the queens and Indian Chiefs the jacks. The stenciled pips are all different colors, the spade suit being blue, the clubs green, the hearts red and the diamonds yellow. While this deck is an important one in the war category, it is also the first known American no-revoke deck, even if not intended as such.

NR3 HUMPHREY REPRINT (not pictured)
Old Sturbridge Village, Sturbridge, Mass. 1976. Although out of chronological order, this deck is listed here, as it is a reprint of NR1 and listed as W2b. It was issued as a Bicentennial deck and sold at the Village.

NR2

NR2 CENTENNIAL PLAYING CARDS
Victor Mauger, New York, 1876 (SX1). This deck, which was issued for the American Centennial, has four different colored suits. The spades are black, hearts are red, diamonds are yellow and clubs are blue.

NR4

NR4 REGULARS
Willis W. Russell, Milltown, NJ, 1906 (refer RU5). This deck was patented with 'Long Distance Pips' with shading in the hearts and spades. The company advertised: "They are a practical solution of the frequent confusion of the suits, particularly at bridge where the dummy is always placed at a distance". The same pips were used on other Russell brands, including Regents and Pinochle.

NR5

NR5 GREEN SPADE
August Petryl, Chicago, 1922. An unusual, beautiful, no-revoke deck. The suit colors are green spades, red hearts, orange diamonds and black clubs. An attempt was made to Americanize the court cards by replacing the kings with Indian Chiefs, the queens with Squaws, and the jacks with Scouts. Different versions of this deck were produced including a Tarock deck of 54 cards plus a Joker and a 78 card Tarot deck. These decks are rare and in high demand.

NR6

NR6 BLUE SPADE PLAYING CARDS

*S.F. Hanzel Card Co., Chicago, 1925. This company produced a series
of four no-revoke decks, of which the Blue Spade is one. Each one
featured blue spades, golden diamonds, green clubs, and red hearts. All
of the decks are identical except for the named ace and the back
pattern. On the pictured deck, the Ace of Spades is the title card, and
the back pattern features blue spades. The other three brands are
Golden Diamond, Green Club and Sweetheart (very rare).*

NR7

NR7 NU FASHION BRIDGE CARDS

*Lefebure Patent Playing Card Co., San Mateo, Calif., 1926. The
design in this deck features white centers on all the pips, including the
indices, of the club and diamond suits. This made the Nu Fashion
cards very efficient for sorting and reading. The Joker gives an example
of each pip.*

NR8

NR8 NEW INDEX SELF SORTING PLAYING CARDS

*USPC., 1926. These cards were another attempt to prevent a revoke.
The only difference between these and a regular deck is in the placing
of the suit signs relative to the indices. The spades and hearts show
the suit symbols above the index number and the clubs and diamonds
are below in the familiar position. These indices were also used on
Bicycle Bridge #86 (NR8a) and at least three other Bicycle Bridge #86
decks with unusual indices (NR8b, NR8c and NR8d), meant to assist
the player in selecting a card from the correct suit when playing Bridge.*

NR9

NR9 NUART

The O.K. Playing Card Co., Tulsa, Oklahoma, 1928. This rare, unusual no-revoke deck has four different colored suits, black spades, red hearts, orange diamonds and purple clubs. The court cards have a Deco feel and feature people in contemporary clothing of the era. The spade and heart kings are dressed in formal attire, while the diamond and club kings wear business suits. The queens are dressed in fashions of the day and the jacks represent different sports.

NR9a

NR9a NUART

The O.K. Playing Card Co., Tulsa, Oklahoma, 1928. This deck is included here, although it is not a no-revoke deck, because it has a similar Ace of Spades and Joker as NR9. The courts are unusual, but representative of a much earlier period.

NR10

NR10 IDEAL NON-REVOKE PLAYING CARDS

George B. Hurd & Co., New York, 1928. A four color deck featuring black spades, red hearts, green diamonds and brown clubs.

NR11

NR11 INNOVATION

Criterion Playing Card Co., New York & Cleveland, 1933. Yet another four color deck trying to cash in on the Bridge craze with black spades, red hearts, green diamonds, and blue clubs.

NR12

NR12 NO-REVOKE

S.R. Huntley, New York, 1932. A four color deck with suit signs slightly modified in shape as well as having black spades, red hearts, orange diamonds and blue clubs. The Ace of Spades features all four colored suit symbols in the large pip. This deck was also used as a bridge teaching deck with lead directions superimposed on the face of the courts and on the aces.

NR13

NR13 AVOID PLAYING CARDS

Avoid Playing Card Co., Tampa, Florida, 1948. Four color playing cards with black spades, red hearts, orange diamonds and purple clubs.

NR14 AVOID PLAYING CARDS II (not pictured)
Avoid Playing Card Co., c1948. The company made a similar deck with gray clubs and pink diamonds.

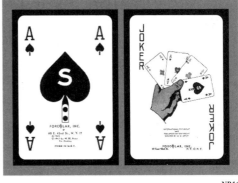

NR16

NR15

NR15 EZ2C PLAYING CARDS
Western Playing Card Co., 1935. The name of this deck is a play on words for 'Easy to See'. They were advertised as the cards with the 'danger signals'. The diamond and club suits have shaded corners highlighting the indices, making this deck almost impossible to sort incorrectly. This concept proved very satisfactory.

NR16 FORCOLAR PLAYING CARDS
Forcolar Inc., New York, 1945. A more modern attempt at a four color 'no-revoke' deck. The colors were black spades, red hearts, orange diamonds and green clubs (in the shape of clovers). The initial of each suit is included on the Ace.

NR17 FORCOLAR II (not pictured)
Forcolar Inc., 1947. An improved version of the above deck with the initial of each suit shown on every pip card, not just the aces.

NEW SUIT SIGNS

Suit signs on playing cards differ throughout the world. In North America, as well as in most English speaking countries, standard decks utilize the original French suit symbols - hearts, spades, diamonds and clubs. Although these symbols are now used universally, many countries still manufacture and use playing cards with their own familiar suits, such as:

Germany - Hearts, Leaves, Acorns and Bells

Switzerland - Flowers, Shields, Acorns and Bells

Italy - Cups, Swords, Batons and Coins

Spain - like Italy but with a different arrangement of the Swords and Clubs.

This chapter is concerned with the early efforts made by manufacturers to modify our familiar suit signs in the hope that card players would welcome the change and that they might profit from obtaining a competitive advantage from the change. Because the first several listings are also included in other chapters, we have only repeated one card from each deck here.

NS3

NS3 ARMY & NAVY PLAYING CARDS
Andrew Dougherty, 1865 (refer W2). Here the new black suits are the Monitors and the Merrimacs, and the red suits, Zouaves and Drummers. This scarce deck is widely sought after by Civil War collectors.

NS4

NS4 NATIONAL CARDS
The Bad Joker, B.W. Hitchcock, NY, c1882 (refer P2). This deck was published using Eagles, Shields, Stars and Flags with a political theme including presidents as the Kings.

NS1

NS1 UNION PLAYING CARDS
1862 (refer W5). This deck was manufactured in a definite hope to take away the hated royalty images and replace them with patriotic symbols.

NS5

NS5 DUTTON'S MILITARY CARDS
c1900 (refer W18). In this deck the suits are the insignia of the different branches of the Army - Cavalry, Infantry, Artillery and Engineers. The cards start from Privates and rise to Four Star Generals.

NS2

NS2 UNION PLAYING CARDS II
1863 (refer W6). Although these new suit signs did not meet with lasting success, some acceptance is evident as this revised version was published a year later.

NS6 MILITAC *(not pictured)*
Parker Brothers, Salem, Mass., 1916. The suit signs in this deck are the same as those in NS5. The Artillery are in red; the Engineers, green; the Cavalry, Yellow; and Infantry, blue. The uniforms are from WW I.

NS7

NS8

NS7 MILITARY FORTUNE TELLERS

H.V. Loring, Chicago, IL, 1918 (refer W27). The new suit signs feature Doves and Bells in the blue suits and Stars and Hearts in the red suits. Soldiers, Nurses and Sailors replace the kings, queens and jacks, and Aviatrixes replace the aces.

NS8 ROODLES

A.J. Patterson, 1912, and later by Flinch Card Co., Kalamazoo, MI. The new suits here are Wishbones, Horseshoes, Shamrocks, and Swastikas. There are 57 cards in this deck, 14 in each suit plus the Joker.

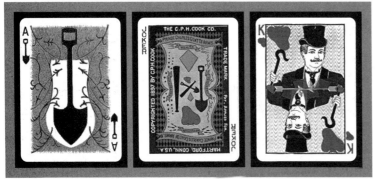

NS9

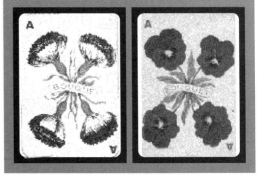

NS10

NS9 PRINCE CHARLES CARDS

C.P.H. Cook Co., Hartford, Conn., 1897. This deck was also known as 'Let's Call a Spade a Spade'. The suits signs use the literal translation of the English meaning of their names. Red hearts are in the shape of a human heart, orange diamonds in the shape of the jewel, blue spades in the shape of a spade or shovel and black clubs in the shape of an actual club or bat. The following quote is from one of three descriptive cards that came with the pack, "The most beautiful card of the whole pack, both in meaning and appearance is the Joker, on which is represented the Prince Charles Coat of Arms, made up of a Heart, Diamond, Spade and Club with their four corresponding maxims - Affection, Wealth, Labor, and Protection, together with every color of the pack represented thereon".

NS10 BOUQUET PLAYING CARDS

Alice D. Ley, Wayne, Nebraska, 1901. The suits in this unusual deck are Violets, Roses, Carnations and Pansies. The court cards are the reigning royalty of the day.

NS11

NS12

NS11 CALENDAR PLAYING CARDS

The Calendar Playing Card Co., Cheyenne, Wyoming, 1925. The new suits in this deck are black Moons, brown Globes, orange Suns and red Stars. There are 13 cards in each suit with the aces numbered 1 as well as 14 and a 'Leap Year' Joker.

NS12 KON QUEST PLAYING CARDS

c1940. The four suits represented here are Moons, Stars, Arrows and Shamrocks.

NS13

NS13 ANMA, ANMA Card Co.
Tulsa, OK, 1941 (refer W35). ANMA is derived from the initials of the four branches of the service - Army, Navy, Marines and Air Corps. The cards rank from the Private, the deuce, to the Colonel as the Ace. The Commander-in-Chief replaces the Joker.

NS16

NS16 INTERNATIONAL PLAYING CARDS
Hiram Jones, 1895. This was the first of many decks issued throughout the years with additional suits. In addition to the standard four, the third black suit is Bullets and the third red suit is Crosses.

NS14

NS14 JAN KEN PO
A.C. Braden for the West Coast Game Co., Pasadena, Cal., 1934. The original Jan Ken Po was a three-suited Japanese game with each suit equal. In Japanese, Jan is Rock, Ken is Scissors and Po is Paper. The original concept was that scissors cut paper, rocks break scissors and paper covers rocks. The addition of the fourth suit Kaji (fire) enabled this deck to be used for all standard games. The rules enclosed with four-suited Jan Ken Po resemble bridge.

NS17

NS17 SECOBRA
The Secobra Inc., 1965. This six-suited deck was designed by Ralph E. Peterson and printed by USPC. The two new suits are Racquets and Wheels which are printed in blue in contrast to the regular red and black suits.

NS15

NS15 RUMME
Milton Bradley, Springfield, MA, 1914. The new suit signs are Crowns, Shields, Helmets and Unicorns. There are 13 cards in each suit plus a Joker.

In the year 1938, contract bridge was in its heyday. Many of the playing card manufacturers hopped on the bandwagon in an attempt to popularize five-suit bridge in order to sell even more decks of cards.

NS18 EAGLE FIVE SUIT BRIDGE
USPC, 1938. The additional suit is
Eagles, printed in green with a circle
around the Eagle suit sign.

NS23 COURT OF MUSIC
Theo. Presser, Philadelphia, c1910. A very interesting deck with suits
of Sharps, Flats, Notes and Rests. The Joker is Pagliacci, the kings and
jacks are composers and the queens opera stars.

NS19 EAGLE FIVE SUIT BRIDGE
Russell PCC, NY. The same deck but the
Eagles are dark green and there is no
circle around the Eagle. This deck was
later reprinted by USPC.

NS20 CASTLE BRIDGE
Parker Bros., Salem, MA, Western PCC.,
1938. The fifth suit of Castles resembles
the rooks from chess and is printed in a
bluish green.

NS21 ARRCO FIVE SUIT BRIDGE
Arrco Playing Card Co., Chicago, 1938.
The fifth suit, printed in green, features
an Eagle holding a globe.

NS24 AMERICA
American Playing Card and Printing Co., Chicago, c1890. This deck
was described as a most interesting and educational social and parlor
game. The new suits are Gold, Grain, Iron and Labor. There are five
face cards - World's Fair, President, Goddess of Liberty, Capitol and
American Flag. The pips run from ace to eight.

NS22 CROMPTON FIVE SUIT BRIDGE (not pictured)
E.E. Fairchild, Rochester, NY, 1938. The additional suit in this deck is
also green and is a combination of the four other suits into one pip.

NS25

NS25 PHILITIS
United States Printing Co., c1890. A rare deck made by a predecessor to USPC. The suits are unusual representations of spades, hearts, diamonds and clubs, likely having something to do with Masonic or other societies.

NS26

NS26 HAND 'EM A LEMON
Lemon Card Co., Milwaukee, c1920. The suits include Berries and Lemons.

COLORFUL AD

ODDITIES

This section focuses on playing cards that were manufactured from unusual materials. Hand made examples, such as the early American Indian cards made from leather are not considered. Odd shapes that were designed for fun, appearance or utilitarian purposes, are also included. New suit signs, discussed in Chapter 31, reflected the constant search for new and better suit symbols, none of which had any lasting success. It will also be seen that, with very few exceptions, the same can be said of these oddities.

O3 ALUMINUM PLAYING CARDS III
Charles L. King, New York, c1920. This was the first real attempt to promote the advantages of aluminum cards over paper cards. The improvement of these cards over the first two listings is that they were printed on enamel that covered the base metal. The cards were packaged in their own attractive aluminum case.

O1 ALUMINUM PLAYING CARDS
Aluminum Mfg. Co., Two Rivers, WI, 1901 (refer SX19). These cards were distributed at the Aluminum Mfg. Co.'s exhibit at the Pan American Expo. Aluminum, although a metal, has the unusual quality of being almost as light as paper. Its most serious fault is that when bent, it is nearly impossible to remove the creases. This deck came in a special aluminum case.

O4 CELLULOID PLAYING CARDS
The Celluloid Co., New York, 1928. These cards were described as permanent. To quote from the box: "Will Last Indefinitely, Washable". This has proven to be false as the card design rubs off with use. In 1928 these cards sold for $3.50 when a deck of ordinary cards sold for between 10¢ to 50¢ per pack.

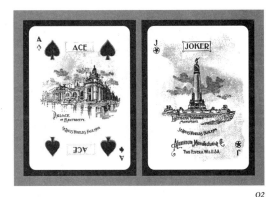

O2 ALUMINUM PLAYING CARDS II
Aluminum Mfg. Co., Two Rivers, Wisconsin, 1904 (refer SX20). A second deck produced with the same purpose in mind. It is difficult to know if this company was trying to promote the metal for use in playing card manufacturing or just trying to promote aluminum. This deck was designed for the Louisiana Purchase Exposition, and sold at the Company's exhibition at the St. Louis World's Fair, 1904. It also came in its own aluminum case.

O4a CELLULOID PLAYING CARDS
Piroxloid Products Corp., New York, 1928. This deck was obviously made by the same company as O4 as it used the same Ace and Joker with only a change in name.

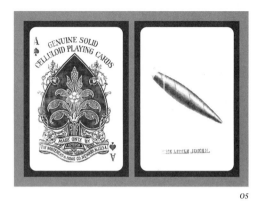

O5

O5 CELLULOID PLAYING CARDS
Whitehead & Hoag, Newark, N. J., c1925. Another set of celluloid cards, this time with a Joker showing a lighted cigar. This was a subtle way to indicate the main drawback of a celluloid deck, the fact that it is highly inflammable.

O7

O7 KLING MAGNETIC STEEL PLAYING CARDS
Regal & Wade, Maspeth, NY. The main purpose for these cards is outdoors play on a windy day. The cards 'cling' to a metal board.

O5a

O5a CELLULOID PLAYING CARDS
Whitehead & Hoag, Newark, N. J., c1925. Another celluloid deck issued by this company has a slightly different Ace of Spades and a more conventional Joker.

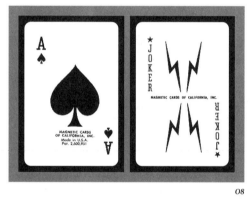

O8

O8 MAGNA CARDS
Magnetic Card Co., California. As with the above deck, these cards are relatively modern. The deck features European style court cards.

O6 ASBESTOS PLAYING CARDS
A. Dougherty, New York, c1925. Apparently people liked the durability of the celluloid cards, but not the combustibility. This deck was made of celluloid, backed with asbestos. It cured the fire problem, but greatly increased the expense. Another fault was that, when stacked, the deck was nearly 2 inches high.

O6

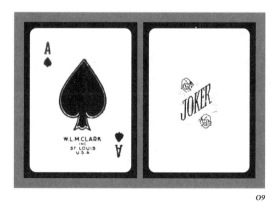

O9

O9 CLARK'S TILES
W.L.M. Clark, Inc., St. Louis, Mo., 1927. This was an earlier version of cards for outdoor play. Their ads read, "Play Bridge in the Breeze or any place you please". The set is based on the Mah Jong idea. It comes with four racks plus 52 small tiles and two Jokers in a handsome wooden box accompanied with instructions.

O10

O10 DUR-O-DECK PLAYING CARD TILES
The Embossing Co., Albany, New York, c1927. A similar deck to O9 deck but not as elaborate. There are no racks, and it was packaged in a plain cardboard box.

O13

O13 GLOBE PLAYING CARDS
Globe Card Co., 68 Cornhill St., c1878. Shortly after I. W. Richardson took charge, the company name was changed to the Globe Card Co. and that name replaced I.W. Richardson on the Ace of Spades. The rest of the deck is identical to O12.

 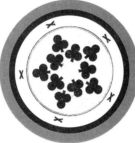

O11

O11 ROUND PLAYING CARDS
I. N. Richardson, 68 Cornhill St., Boston, Mass. These were patented in 1874 and although round cards have never really enjoyed great popularity, they have been made for a long time. They were first produced in Germany in the 15th century and are still being made throughout the world today.

O11a ROUND PLAYING CARDS *(not pictured)*
I. N. Richardson, 68 Cornhill St., Boston, Mass., c1875. I. N. Richardson produced a second round deck with several improvements, the most important being different color suits - black spades, red hearts, green clubs and yellow diamonds.

O14

O14 GLOBE PLAYING CARDS
Globe Card Co., 78 Hawley St., Boston, c1880. The company moved to larger quarters, but continued to produce the same deck. The new address appears on the Ace of Spades.

O12

O12 ROUND PLAYING CARDS
I. W. Richardson (son of I. N.), 68 Cornhill St., Boston, c1875. These cards also had the four different color suits and a further improvement, the king's crown, queen's hair and jack's cap featured the suit's color. The Joker pictured was the same for each of the five Richardson and Globe decks.

O15

O15 WATERPROOF PLAYING CARDS
The Waterproof Playing Card Co., 50 Bromfield St., Boston, c1878. This round deck had newly designed courts and used only red and black suit symbols. It is hard to understand the term 'waterproof', as the cards were certainly not washable. In the original Encyclopedia it mentioned that there were two Jokers that appeared identical except that one read "The Globe". The reason for this has recently come to light thanks to the research activities of Michael Goodall who unearthed an article from an 1870s Stationers Magazine describing a lawsuit heard in Boston. The complainants Lee & Shepard, were successful in prohibiting the defendants, W. F. Miller and F. D. Montgomery from using the word "Globe" in connection with their 'Waterproof Playing Cards' (see also O28).

O16

O20

O16 INTERNATIONAL ROUND CARDS

The International Playing Card Co. of America, Chicago, c1930. This is a truly beautiful deck of round cards with artistic two headed courts. Each suit is a different color - black spades, red hearts, green clubs and blue diamonds. Their advertising demonstrated good reasons for using round cards. To quote: "The new round card fits the hand better, does not break or bend, and shuffles easily. Any card can be seen without confusion as to suit. The marks are clockwise around the entire circle and can be seen from side, bottom or top. Round cards cannot be marked as they do not assume the same position twice alike after shuffling. There are no corners to break off, which often ruins an entire deck." This deck was used for an advertising deck by several firms, especially the International Pressmen Union who were likely connected with the maker.

O20 TRANSPARENT PLAYING CARDS

M. Nelson, New York, c1860. This was the same Nelson who published the two famous Civil War Generals cards and other Novelty decks. They made a series of transparent decks, all with the same court cards but with different subjects revealed when held to the light. Among these were: Illustrations from five plays by Shakespeare, Women of Ancient and Modern Times and Problem Caricature pictures (refer also to N48a). It appears that the early Nelson decks were transparent (with no Nelson identification) and that Nelson subsequently came out with new improved cards - the ones with the picture on the face of the card and a tiny, Triplicate-like card image in the top left corner (refer W7, W8, W42, N48 and N48a).

O17

O17 CIRCULAR COON CARDS

Sutherland Co., New York, 1925. An unusual round deck with three-headed caricature style courts, all smiling. The back features a black man dressed in flamboyant clothing

O18

O18 DISCUS ROUND CARDS

Arrow Playing Card Co., Chicago, c1930. These cards feature four headed courts.

O20a

O20a TRANSPARENT PLAYING CARDS

Michauds Manufacturers, c1860. This deck has scenes identical to some of those on Nelson's Love Scenes and others from his Shakespeare deck. Was Michauds from France? Did Nelson copy his decks? Or, was Michauds an American maker (the quotations on each card are in English) copying Nelson?

O19

O19 ARRCO ROUND CARDS

ARRCO, Chicago, c1950. The same deck as O18 but with the new company name.

O20b

O20b TRANSPARENT PLAYING CARDS

New York, c1870. The Ace of Spades of this deck is almost the same as the Nelson transparent but with the Nelson name 'erased' from the banner on the Ace. However the courts are newer and double-headed but without indices. The scenes are from MacBeth and other Shakespeare plays.

O20c

O20c TRANSPARENT PLAYING CARDS
New York, c1860. A different transparent deck that is not identified anywhere on the deck. The cards were perhaps made for the 'southern' market as a number are pro-slavery and derogatory towards blacks.

O24

O24 NEW ERA CONCAVE PLAYING CARDS
Standard Playing Card Co., Chicago, Ill., 1929. The shape is unusual and the deck is difficult to handle.

O21

O21 TRANSPARENT PLAYING CARDS
The Transparent Playing Card Co., New York, c1880. These are standard playing cards which when held to the light reveal hidden pictures on all the cards except the courts. The same Joker was used for the two subsequent listings.

O25

O25 E.Z. PLAYING CARDS
E.Z. Playing Card Co., Quincy, Mass., 1952. This oddly shaped rectangular deck was designed so that one could hold the cards comfortably. The smaller index in the upper right hand corners is to prevent one from pulling out the wrong card.

O22   *O23*

O22 TRANSPARENT PLAYING CARDS
The American Playing Card Co., New York, c1880. The same deck as O21 but with a different Ace of Spades and a different set of hidden pictures.

O23 TRANSPARENT PLAYING CARDS
c1880. When comparing this deck with O21 and O22, it seems clear that the same company made all three decks.

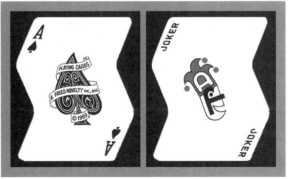

O26

O26 THE CROOKED DECK
A. Freed Novelty Co., New York, 1969. An odd deck manufactured for laughs rather than play.

O27

O30

O30 POPULAR PLAYING CARDS
J.H. Bufford's Sons, Boston, Mass., 1877. This is the only American transparent deck which has a back design. It is still relatively easy to view the hidden pictures.

O27 CONTOURA
USPC, c1940. A barrel shaped or convex deck with no practical advantages was introduced by USPC.

O31

O31 GRANDMOTHER STOVER'S MINIATURES
Grandmother Stovers Inc., Columbus, Ohio. A very tiny deck, 8 x 10mm, which was advertised as the smallest deck ever made. It came with a tiny booklet entitled "How to Play Cards". Needless to say the entire package is too small to be used.

O28

O28 GLOBE PLAYING CARDS
Miller & Montgomery, Boston, c1877. This deck is the version of Waterproof Playing Cards (refer to lawsuit under O15) with the Joker that says "The Globe".

O32

O29

O29 GLOBE PLAYING CARDS
The Globe Playing Card Co., Boston, c1890. This deck is completely different from O13 and O14. The suits are printed in black on colored backgrounds – spades on lilac, hearts on orange, diamonds on yellow and clubs on green. The deck comes packaged in a round, Bakelite box decorated with their Ace of Spades.

O32 TINY PLAYING CARDS
Theodore L. DeLand, Philadelphia, c1910. Actually O31 was not the smallest; this honor belongs to Deland, best known for his magic cards. His tiny deck, 6 1/2 x 9mm features the 'Baby' Joker. The box announced "The Commissioner of Internal Revenue has decided that these cards are exempt from taxes".

NOVELTY PLAYING CARDS

Novelty playing cards and new innovations are the subjects covered in this chapter. These feature non-standard decks with special subjects and/or novel courts. It does not include decks that have changes in their court cards made by playing card companies in an effort to improve their decks, and then sold as part of their regular line.

No advertising decks are included here. We have also omitted non-standard decks that fit into the other chapters, such as Transformation, Souvenir, World's Fair, etc. In addition we have ignored decks issued since WWII unless listed in the original Encyclopedia.

N2a DELAND'S AUTOMATIC DECK
Theodore L. DeLand, Philadelphia, 1913. One of the earliest DeLand decks was also one of his most popular. It was marketed as DeLand's Dollar Deck.

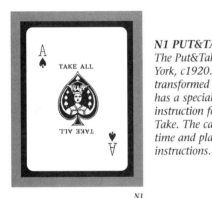

N1 PUT&TAKE PLAYING CARDS
The Put&Take Playing Card Co., New York, c1920. A standard deck of cards transformed for double duty. Each card has a special overprint with some instruction for the game of Put and Take. The cards are picked one at a time and players must follow instructions.

N3 DELAND'S NIFTY
S. S. Adams, Asbury Park, 1915 to 1940. All early issues of the Adams decks have their names on the Ace of Spades. In 1925 these names were dropped and all of the decks now used the same 'Baseball Ace' and the N2 Joker pictured.

N2 DELAND'S AUTOMATIC PLAYING CARDS
S.S. Adams, Asbury Park, N.J., 1918. Theodore L. DeLand sold his card business to the S. S. Adams Co. in 1918. This was one of the first decks re-published by Adams. As with all the DeLand/Adams decks, the backs were printed in such a way that one who knew the key could ascertain the suit and rank of every card. Adams completely redesigned the courts and Joker for their versions.

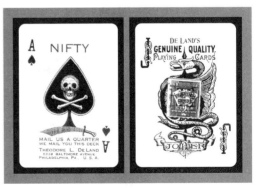

N3a DELAND'S NIFTY DECK
Theodore L. DeLand, Philadelphia, 1913. A rare example of another of DeLand's early decks has a skull and crossbones Ace of Spades and a Joker reminiscent of early NYCC Best Bowers. The markings on this deck were inconspicuously placed on the edges of the cards. The original tax stamp can be seen covering the middle of the pictured Joker.

N4 DELAND'S DAISY DECK

S. S. Adams, Asbury Park, 1918 to 1940. This is the same basic deck as N3 but the back features a different marked design. The extra card shown explained the markings.

N8 BUSTER BROWN PLAYING CARDS

USPC, 1906. Buster Brown was a popular comic strip character and all of the 52 cards and the Joker feature a series of cartoons about him and his dog Tige. This miniature patience deck measures 1 3/4 x 2 1/2 inches.

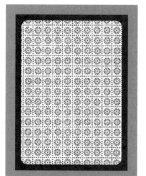

N5 DELAND'S STAR DECK

S. S. Adams, Asbury Park, 1918 to 1940.

N6 ADAMS' LEAGUE DECK

S. S. Adams, Asbury Park, 1918 to 1940. This deck has a baseball theme marked back.

N9 TEDDY BEAR PLAYING CARDS

Standard Playing Card Co., Chicago, 1907. Another cartoon patience deck, the same size as N8. This deck had a standard Joker and standard courts.

N7 SUPER PLAYING CARDS

Knapp Electric Co., Indianapolis, Indiana, 1931. "A Master Magic Creation" similar to the above five decks. The back designs of all of the marked decks were patented.

N10 THE BOOB DECK

c1930. This patience deck features comical court cards, as well as witty sayings on the aces and pips. Each suit repeated the witticisms.

N11

N11 CARNIVAL CARDS
The Carnival Playing Card Co., New Orleans, 1925. Copyright by Harry D. Wallace, New Orleans. The Ace of Spades reads Mardi Gras. The court cards are artistic, with each of the four suits representing a different Mardi Gras figure. Spades feature Proteus, a sea God, who had the power to change his form at will. Hearts represent Comus, the God of festivity. Clubs are Momus, the God of mockery. Diamonds are Rex, the King of the Mardi Gras. This deck was reissued in 1981 by the Historic New Orleans Collection with this information clearly printed on the Ace of Spades.

N12

N12 MARDI GRAS CARDS
George F. Castleton, printed by Tropical Printing, New Orleans, 1930. An unusual and interesting deck featuring special courts representing figures from the Mardi Gras and special aces showing views of the City of New Orleans.

N13

N13 PEPPER PLAYING CARDS
E.W. McCarroll, Pittsburgh, Penn.,1925. Each card in this unusual deck has two sets of indices. It can therefore be used as a regular deck by ignoring the bottom set, or used to play versions of certain games such as Poker using both sets. It comes with an instruction card and leaflet.

N14

N14 DOUBLE ACTION PLAYING CARDS
The Double Action Playing Card Co., NY, 1935. This deck, with a similar purpose to N13, is divided diagonally in half. One side has a blue background and the other is white. Excerpts from the instruction card include: "The cards are assorted in zones, giving each player 2 hands (one in white and the other blue). Any regular game can be played, by confining use to either zone. Variations in games: in Bridge, the winning bidder can name the zone to be used; in draw poker, draw to either zone, and after the draw, examine both hands and play the better". There are two complete decks, but not identical, as the White Ace of Spades shares with the Blue 9 of Hearts, while the Blue Ace of Spades has the White 5 of Diamonds.

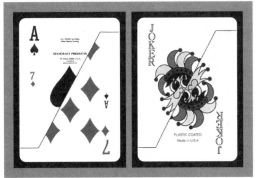

N14a

N14a DOUBLE ACTION PLAYING CARDS
Stancraft Products, c1950. Another deck with the same idea was produced by Stancraft.

N15

N15 DUPLEX DECK
The Duplex Playing Card Co., 1929. This is the same type of deck as the last two where each card has two sets of values. This deck also has a complete set of dominos on the face of the cards. In order to complete the 55 piece domino set, the deck contains three Jokers.

N16   N17

N20

N16 CHESS CARDS

c1900. This set of chess cards (maker unknown) is included here as the cards have suit marks and can be used in standard card games. The suits are at the bottom of the chess symbols, both in the card and indices. This deck is 32 cards to match up with the 32 chess pieces.

N17 DICE PLAYING CARDS

c1930. This is still another double duty deck. Each card, with the exception of the Kings (two dice adding to seven), has the face of a single die. The aces have a single spot, the two's have two spots, etc. The sequence is repeated starting at card number seven.

N20 INCA

(Good Neighbor), Brown & Bigelow, 1948. This deck was made for the American Association of University Women. The courts are done in the art style of the Inca Indians of Peru. They are beautifully designed and colored. They were originally sold in a double box with N21.

N18

N21

N18 VANITY PLAYING CARDS

The Eckley Sales Co., San Francisco, CA, 1928. This deck features specially designed courts with a modernistic feel. All the suit symbols are modified as to shape, giving them a squared look.

N21 MAYA

(Good Neighbor), Brown & Bigelow, 1948. The mate to N20 has the style of the Mayans.

N19

N22

N19 HYCREST MODERN ROYALTY

Hycrest Playing Card Co., New York, c1931. The deck has very different modern courts in beautiful color combinations and a striking Joker. These, combined with new enlarged indices, create a startling effect. This is a scarce and very collectible deck.

N22 PHOTO PLAYING CARDS

The Photo Playing Card Co., Fresno, California, c1930. An unusual deck of standard playing cards, available for sale with your family photograph or individual snapshot on the backs. The regular and advertising Aces of Spades are shown.

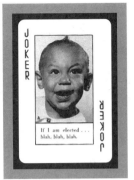

N23

N23 BANNISTER BABIES I
Brown & Bigelow, 1954. Humorous baby pictures presented with a political theme (refer P7).

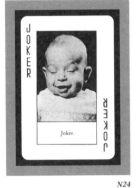

N24

N24 BANNISTER BABIES II
Brown & Bigelow, 1954. This pack was issued about the same time as N23, but the captions on a different set of photographs relate to card playing.

N25

N25 MONKEY CAPERS
Brown & Bigelow, 1958. A deck utilizing the same captions as N24 was produced with photographs of monkeys and chimpanzees.

N26

N26 BELIEVE IT OR NOT
Creative Playing Card Co., St. Louis, Missouri, 1963. 53 cartoons by Robert Ripley with facts from his 'Believe it or Not' books.

N27

N27 TEE UP
Creative Playing Card Co., St. Louis, 1962. The first of a series of six decks each featuring a set of 54 cartoons relating to a particular activity. Tee Up attempts to bring out the humorous side of golfing.

N28   N29

N30   N31

N32

N28 FISH UP, 1963.

N29 BOWL UP, 1963.

N30 CHEER UP, 1963.

N31 DRINK UP, 1963.

N32 TUNE UP, 1963.

N33

N33 STAG PARTY PACK

1953. A deck composed of 53 cartoons. Unlike the 'UP' cards, these cartoons are suggestive and occasionally naughty.

N33a

N33a STAG PARTY PACK

1953. Another deck composed of 53 cartoons.

N33b

N33b STAG PARTY PACK

1953. Another naughty cartoon deck was produced about the same time.

N35

N35 VARGAS VANITIES

Western World Playing Card Co., St. Louis, 1953. This deck features 53 reprints of paintings of beautiful women by Alberto Vargas. A second Joker gives the history of this artist famous for his 'Esquire Girls'.

N36   N37

N36 ART BALL

Don Celender, 1972. An unusual pack, produced for the Greenwich Village Art Colony. In this deck, the heads of contemporary artists are superimposed onto the bodies of some National Football League players.

N37 MUSEUM OF MODERN ART DECK

USPC, 1972. This deck was produced for the New York Museum and designed by Bruce Blackburn. The background for each suit is a different color. Spades are blue, hearts are red, diamonds are orange and clubs are green. All the suit signs are white and the ranks are printed in black.

N34

N34 PEPPY

United States Inc., St. Louis, Missouri, 1954. This deck is the same type of deck as N33 with two Jokers and advertised as naughty but nice comic cards "for men only".

N41 SURVIVAL PLAYING CARDS
Environs, Inc., Hood River, Oregon, 1974. This deck was designed to help campers and hikers cope with emergency situations. The spade suit deals with getting lost and finding food and water. The hearts relate to first aid for bleeding, shock or breathing difficulties. Diamonds are for wounds, burns or bites and the clubs handle shelter and signaling for help.

N38 ESKA CALORIE COUNTER DECK
ARRCO, Chicago, for the Smith, Kline & French Laboratories. Each card features a tempting, non-diet dish and shows the number of calories. The same items are repeated in each suit.

N39 TRIP TRAP
Stancraft, St. Paul, Minn., 1970. Designed by W. R. Spence, MD of Spenco, the deck tries to combat drug abuse. Each of the pip cards describes a different narcotic, the slang name, medical uses (if any) and the symptoms displayed by an abuser. The court cards show warnings to potential users. One of the Jokers lists the drugs and the other features a message from Dr. Spence.

N42 GREEN CROSS FOR SAFETY
Brown & Bigelow for the National Safety Council, 1954. Each card features a different suggestion for preventing accidents at home and at work. Including the two Jokers, there are 54 safety hints.

N40 ROYAL FLASH
Du-Rite Enterprises, 1974. This deck might have been included under new suit signs, but the intent was not to create a new format. As if to combat N39, this deck has as its suit symbols, plants used in the manufacture of narcotics, among which are poppies, hemp, marijuana, etc.

N43 O-SHLEMIEL
Arrco, Chicago, for Originals Only Company, NY, 1976. A novel deck with Yiddish expressions and their English translations in the center of each card. They can be used for regular games, or the game of 'Tsuris' (trouble), when using the letter and number symbols.

N44

N44 SHEBA PLAYING CARDS

Sheba Products, a division of Omega Concepts Ltd., New York, 1972. This deck is dedicated to the glory of the African Peoples. The kings represent Askia, the Great, Emperor of Songhay, 1493-1539. The queens feature Makida, the Queen of Sheba and Ethiopia, c1000-950 BC and the jacks represent Hannibal, the Carthaginian General, famous for crossing the Alps.

N45

N45 SOUL CARDS

Soul-Mar Inc.,Whittier, Cal, 1973. This deck is "dedicated to Black People everywhere".

N46

N46 FACT AND FANCY

Designed by Dick Martin for the Chicago Playing Card Collectors, on their 10th Anniversary, 1961. Every card features a historic pack or card player from yesteryear. There are 52 cards, a Joker, and a title card which reads "A Deck of Decks and Playing Card Curiosa". It also shows the number of their limited edition of 600 copies. It comes with a booklet describing "Six Centuries of Famous Card Players and Playing Cards".

N47

N47 THE MYTHOLOGICAL ZOO

Designed by Dick Martin for the Chicago Playing Card Collectors' 20th Anniversary, 1971. Each card features a drawing of a mythological creature. This deck includes a Joker, a title card and a booklet that describes each card. There were 300 copies made. The artist hand-colored 100 decks, which were numbered and signed. The uncolored version came in a wrapper, while the colored one was boxed.

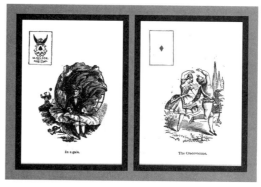

N48

N48 LOVE SCENES

M. Nelson Co., 1864. As well as his scarce transparent decks (O20), Nelson produced several other decks in the same style as Love Scenes. It appears that the early Nelson decks were transparent (with no Nelson identification) and that he subsequently came out with new improved cards - the ones with the picture on the face of the card and a tiny, Triplicate-like card image in the top left corner (refer also W7, W8, W42 and N48a). This deck has a miniature card in each corner identical to those on the Confederate and Union Generals decks.

N48a

N48a WAR SCENES

M. Nelson Co., 1864. A recent discovery, this deck has 'humorous' scenes on each card with an unusual back done in green ink. The scenes are comic and mainly related to the Civil War times.

N49

N50

N53

N49 CHECK YOUR CATCH

Fin & Feather Playing Card Co., Petoria, Illinois. Each of the 52 cards and the two Jokers features a color sketch of an American fish. The deck comes in leather bag with the title of the deck on the front.

N50 AUTHORS

J.H. Singer, New York, 1888. This is the game of Authors as well as a 52 card deck. It states that "any game can be played that is also a full deck of 52 cards". There is no maker's name on the cards or box.

N53 IMPERIAL AUTOMOBILE CO.

Boston, c1910. This appears to be an advertising deck with a 'rooter' theme. All the courts are represented by 'pig people'.

N51

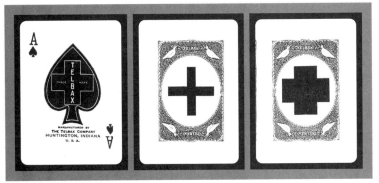

N54

N51 CUDAHY BAR S

Cudahy Bar S Meat Packing Co., Phoenix, Arizona, 1974. Ed Mell designed this deck. The court cards repeat in each suit and each pip card has biographical notes and a sketch of a famous personage of the old West.

N54 TELBAX

The Telbax Company, Huntington, Indiana, c1920. Another deck with a 'marked' back. The name says it all; you can tell the denomination of each card from the different size cross on the back.

N52

N52 McLOUGHLIN BROS.

New York, c1895. This patience size deck was made by the well known manufacturer of children's games and books; perhaps inserted in a box for a game requiring playing cards.

N55

N55 MIDGET

Peter G. Thompson, Cincinnati, c1890. This miniature deck has double-ended cards with no indices, indicating a date in the 19th century.

INDEX

Rhineland Pinochle, 138

Rhode Island Souvenir, 260

Ricco Steamboat, 147

Rice, J., 10

Richardson Ball Bearing Skate Co., 186

Richardson, I.N., 297

Richardson, I.W., 297

Ride and Win, 184

Rider-Waite Tarot, 241

Rieder, M., 255, 256

Ritz Carrollton, 148

Rivals #1110, 44

Rivals #15, 40, 44

Riviera, 138

Road Transportation, 195

Robertson, James, 10

Robitaille, S., 168

Rock Island Souvenir, 268

Rocky Mountain Souvenir, 260

Rogers, J. L. & B. A., 282

Rogue #831, 173

Roll-Crawford-Brendamour Company, 199

Romo & Kredi, 272

Roodles, 290

Rooster #38, 120

Roosters #100, 117, 118

Root Playing Card Co., 224

Rough Backs, 33, 34

Rough Rider, 150

Rough Riders #90, 29

Roulette Cards, 127

Round, 297

Round Cornered Whist, 46

Roundup #902, 137, 138

Rouser #549, 136

Rovers #20, 40

Royal (W, B & R), 177

Royal, 247

Royal Bridge, 168

Royal Flash, 307

Royal Flush, 27

Royal Playing Cards, 232

Royal Revelers, 228

Ruby Queen, 59

Rugby, 138

Ruledge Playing Card Co., 278

Rumme, 291

Russell & Morgan (R&M), 3, 17, 81, 82, 83, 84, 85, 86, 88, 92, 109, 125, 167, 277

Russell & Morgan Printing (RMP), 81, 85, 87, 89, 90, 91, 92, 93, 94, 95, 96, 103

Russell Playing Card Co., 4, 31, 38, 121, 125-136, 142, 143, 145, 163, 280, 292

Russell's Blue Ribbon #323, 127

Russell's Club Mogul, 126

Russell's Mogul, 126

Russell's Pinochle, 126

Russell's Recruits, 125

Russell's Regents, 125

Russell's Regulars, 126, 220

Russell's Regulars Skat, 126

Russell's Retrievers #40, 127

Russell's Rustlers, 126

Rustlers #136, 131

Rustlers Pinochle #118, 131

Ryves & Ashmead, 10

Ryves, Edward, 10

S

S.F. Hanzel Card Co., 286

S.H. Hines Co., 202

S.O. Barnum & Son, 255

Saladee's Patent, 27, 56

Saml. Cupples & Co., 24

Saml. M. Stewart, 10

Samuel Cupples Envelope Co., 250

Samuel Hart & Co., 3, 13, 14, 27 45, 49, 51, 53, 54, 55, 56, 58, 66, 68, 206, 227

Samuel Hart Transformation, 206

Samuel M. Stewart & Co., 20

Santa Anita Ranch (Lucky Baldwin), 264

Saratoga #11, 138

Sauzade, Robert, 10, 21

Scamper, 160

Schering Medical, 183, 184

Schlitz, 187

Scottie, 157

Seaboard Railway Souvenir, 268

Sebago, 78

Secobra, 291

Self-Sorting, 286

Seminole Wars, 19, 217, 285

Senate, 106, 146

Senators #40, 41

Service, 152

Sesquicentennial, 252

Shakespearian Playing Card Co., 60

Sheba, 308

Sheffer, Max. B., 149

Sheldon's Fortune Telling, 246

Shirley Temple, 240

Shuffler #744, 170

Silhouette, 158

Silk Velour, 42

Silver City Playing Cards, 117

Singer, J.H., 246, 309

Skat #2 German Faces, 94

Skat #29, 58

Skat #38, 72

Skat #4 American Faces, 94

Skyways Pipe Tobacco, 216

Smart Set #400, 129, 131, 163

Smith, L. Lum, 231

Snipe Plug Cut, 210

Snowshoes, 168

Socialist Playing Cards, 232

Socials #36, 32

Society #1000, 137, 138

Society #1001, 138, 163

Society Pinochle, 138

Sodium Bicarbonate, 186

Solo #21, 41

Solo #36, 72

Solo #75, 32

Solo, 46, 117

Solo Pinochle, 37

Solo, Euchre and Pinochle, 117

Soo Line, 268

Soul, 308

South Africa, 272

Southern Pacific Lines Souvenir, 267

Souvenir Card Co., 261

Souvenir Playing Card Co., 256

T

X

Y

Z